AMANDA SPARKMAN

# POLITICAL NUMERACY

# POLITICAL NUMERACY

## Mathematical Perspectives on Our Chaotic Constitution

**BY MICHAEL I. MEYERSON**

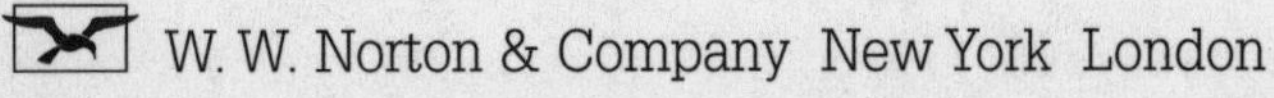
W. W. Norton & Company New York London

Copyright © 2002 by Michael I. Meyerson

All rights reserved
Printed in the United States of America
First published as a Norton paperback 2003

Since this page cannot legibly accommodate all the copyright notices, pages 277–78 constitute an extension of the copyright page.

For information about permission to reproduce selections from this book, write to Permissions, W. W. Norton & Company, Inc., 500 Fifth Avenue, New York, NY 10110

The text of this book is composed in 11 point Granjon with display set in Helvetica Inserat. Composition by Gina Webster. Manufacturing by Quebecor World Fairfield, Inc. Book design by Dana Sloan. Production manager: Andrew Marasia.

Library of Congress Cataloging-in-Publication Data
Meyerson, Michael.
Political numeracy : mathematical perspectives on our chaotic constitution / by Michael I. Meyerson.
p. cm.
Includes bibliographical references and index.
ISBN 0-393-04172-7
1. Mathematics—Social aspects. 2. United States—Politics and government. I. Title.

QA10.7 .M48 2002
510—dc21 2001044818

**ISBN 0-393-32372-2 pbk.**

W. W. Norton & Company, Inc., 500 Fifth Avenue, New York, NY 10110
www.wwnorton.com

W. W. Norton & Company, Ltd., Castle House, 75/76 Wells Street, London WIT 3QT

1 2 3 4 5 6 7 8 9 0

*To Lesly, William, and Andrew. With Love.*

# Contents

# Acknowledgments

I am indebted to many people who have helped me with this book. First, I would like to thank F. Michael Higginbotham, Charles Tiefer, Roberto Allen, Cathy Jones, and William Meyerson for their helpful suggestions and criticisms. The errors, misstatements, and problems that remain, of course, are solely my own.

My student researchers, Roger Wolfe, Rebecca L. Roberts, and Catherine Sauvain, provided invaluable assistance. John Sebert, former dean of the University of Baltimore School of Law, always offered encouragement during the gestation period of this project. Thanks also go to the current University of Baltimore School of Law dean, Gilbert Holmes, for his support as this project neared its completion.

I owe an enormous debt to the superb research staff at the Law Library of the University of Baltimore, led by Emily Greenberg. Deepest gratitude goes to Elizabeth Rhodes, reference and faculty liaison, for her creative research assistance and unfailing support.

The people at Norton have been a joy to work with. I am

eternally indebted to my editor, Alane Mason, for having both the courage to accept this book and the insight to improve it exponentially. I would also like to thank her assistant Stefanie Diaz, for being a pleasure to work with through a long and arduous process.

My agent, Geri Thoma, has been a source of great strength and comfort. I would also like to acknowledge the late Clyde Taylor. He was my first agent, and believed in the book before almost anyone else. He is sorely missed.

My parents, Jack and Marian Meyerson, taught me the beauty of mathematics and the importance of justice. I hope this book does them justice.

Finally, I would like to thank my wife, Lesly Berger, and my children, William and Andrew. They helped in countless ways, from reading the manuscript to arguing over politics to editing and reediting the work. Most important, though, even before this book was begun, and through all the research, writing, and rewriting, they have made my life worth living.

# Preface

"If you were any good in math, you'd all be in medical school." Thus spoke my torts professor to his class of eager first-year law students, who did not disagree.

The concept that mathematics can be relevant to the study of law seems foreign to many modern legal minds. Certainly, the presidential election of 2000 forced a discussion of the intricacies and merits of the electoral college. Statistical analysis appears in the courtroom and at agency hearings. Those in the law and economics movement use mathematics in their analyses, as do the social choice theorists. Yet, in all these cases, mathematics, in the derisive words of one of my colleagues, is "just a tool." And to many others, the absence of mathematics is one of law's greatest appeals.

The current gulf between mathematics and law is not surprising. As studied in school, mathematics can be mechanical, repetitive, and often downright boring. But mathematics is far more than rote learning and intricate formulas. When told that one of his students had decided to drop mathematics to become a poet, David Hilbert, one of the leading mathematicians of the twentieth century, replied, "Good—he did not have enough imagination to be a mathematician."

Mathematics should not be confined to the material needed to pass a class in trigonometry, geometry, or calculus. Lewis Carroll was much closer to the truth than many realize when he identified "[t]he different branches of Arithmetic—Ambition, Distraction, Uglification, and Derision."

There is, to the astonishment of many nonmathematicians, an aesthetic quality to mathematics, which often transcends intuition, and even imagination. At its heart, "mathematics is the science of patterns." It identifies relationships and, more importantly, teaches how to identify patterns that might be far from obvious. "If mathematics is anything, it is the art of choosing the most *elegant* generalization for some abstract pattern."

Intelligent thinking in general and legal thinking in particular require just such a skill. Seeing relationships between doctrines, cases, or provisions of the Constitution is essential for establishing a coherent way of approaching legal issues. There is much that mathematics can teach about recognizing legal patterns.

The great U.S. jurist Oliver Wendell Holmes Jr. is widely quoted as saying, "A page of history is worth a volume of logic." Holmes knew that mathematical logic could never be the exclusive basis for legal reasoning. Yet Holmes was also a student of mathematics who understood that mathematics can explain concepts at the heart of all reasoning. He was taught this lesson at an early age by his father, Oliver Wendell Holmes Sr. In an immensely popular series of pre-Civil War essays, *The Autocrat at the Breakfast Table*, the senior Holmes explained the need to reason abstractly, in the universal rather than in the particular: "All economical and practical wisdom is an extension of the following arithmetical formula: $2 + 2 = 4$. Every philosophical proposition has the character of the expression $a + b = c$. We are mere operatives, empirics, and egotists until we learn to think in letters instead of figures."

Thomas Jefferson, too, was a superior mathematician. He was quite comfortable with the then relatively recently invented techniques of calculus. He was able, as he wrote a friend, to design a plow, using the "*Doctrine of Fluxions* . . . [to calculate the] form offering least resistance to the rising sod." His very thinking process was permeated with mathematical considerations. "Jefferson's intellectual world and his daily life were regulated by numbers to a degree that seems astonishing to a reader in the twentieth century. He was skilled in mathematics and delighted in numbers and in calculation. Almost every aspect of his life was reduced to numerical observations and calculations." For instance, to allay fears that Shays's Rebellion indicated a fundamental weakness in the great American experiment, Jefferson calculated that, statistically speaking, this single flare-up was not a cause for concern. "[O]ne rebellion in 13 states in the course of 11 years," he wrote, is equivalent to "one for each state in a century and a half."

Retired from the presidency, Jefferson turned even more of his attention to the study of mathematics. He wrote to John Adams that he had "given up newspapers . . . for Newton and Euclid."

It is important to realize the reasoning behind the framers' selection of two-thirds as the fraction for veto overrides and treaties, as well as why we have two different houses of Congress and 535 presidential electors. But significant other insights arise from nonarithmetic areas of math.

The very process of legal reasoning is closely tied to the Euclidean system of logical reasoning; "deducing conclusions from axioms about undefined terms . . . [is part of] the typical process of lawyers." Those interpreting the Constitution, from

Alexander Hamilton to modern Supreme Court decisions, have utilized this form of proof.

Of course, legal reasoning is far more subjective and political than a geometric proof, but those who have challenged a dry, mechanical vision of law are part of the mathematical tradition as well. The innovations of non-Euclidean geometry in the nineteenth century resulted in a revolution in thinking which toppled an intellectual edifice that had stood for more than 2,000 years. The study of non-Euclidean geometry reveals that, depending on initial assumptions, the same issue can be addressed legitimately in seemingly contradictory ways.

Oliver Wendell Holmes Jr. was "said to have invented non-Euclidean legal thinking" by challenging the orthodoxy of legal reasoning. Rather than merely accepting fundamental principles, or axioms, as immutable truths, both the non-Euclidean mathematicians and modern legal reasoners have recognized that alternative principles are possible. Mathematical and legal "[a]xioms have been secularized. They are now regarded merely as assumptions, and no assumptions are considered sacrosanct."

An understanding of mathematics frees rather than imprisons the mind. It is not a matter of reducing values, theories, or choices to numbers; rather it requires us to admit that most things in life cannot be quantified. And fancy graphs do not prove the inevitability of a legal concept. As has been said before, "Any idiot can draw a graph."

A fuller understanding of all that modern mathematics has to offer will make for more sophisticated constitutional reasoning. Modern math makes it easier to see that problems may be difficult, not because of lack of knowledge but because of their inherent nature. The study of infinity reveals contradictions that mirror the intricacy of dealing with competing rights of infinite importance. The paradoxes that led to Kurt Gödel's incompleteness theorem have close analogues to self-referential constitutional issues, such as whether the executive branch can be trusted to investigate allegations of presidential wrongdoing. Chaos theory reveals why predicting the path of Supreme Court decisions may be a futile venture.

Other areas of mathematics can create pictures that permit more creative views of constitutional jurisprudence. Topology, which explores nonrigid shapes, provides an interesting framework for considering the changeable relationship between the federal and state governments, a configuration which is flexible but not unbounded. And game theory presents new ways of thinking about conflict and cooperation within our constitutional system.

This book represents the beginning of a rapprochement between two fields that have drifted apart. But it is only a beginning. There is much more to be said about the mathematical topics discussed, and there are many other areas of mathematics that were not included at all. Moreover, many of the insights presented undoubtedly will be revised or rejected. My review of mathematical reasoning has reaffirmed for me the necessity of modesty in announcing one's conclusion. John Allen Paulos, who has led the battle to end "innumeracy," has noted that "people often charge that a knowledge of mathematics leads to the illusion of certainty and a consequent arrogance. I think this is false. . . ."

Rather than leading to certainty, an understanding of mathematics can facilitate a greater respect for differing constitutional perspectives, especially those with which we most strongly disagree. This respect, I hope, will lead to a greater willingness to listen and a stronger sense of humility in one's own conclusions.

## Introduction

# The Ugliest Number in the Constitution

Beauty may be in the eye of the beholder, but ugliness is sometimes indisputable. There is no number, in fact no concept, in the entire Constitution as hideous as three-fifths, the fraction by which slaves were arithmetically devalued as people.

Known as the three-fifths compromise, Article I, Section 2, of the Constitution decreed that both representation and direct taxation be apportioned among the states "according to their respective Numbers." Each state's "Number" was to be calculated "by adding to the whole Number of free Persons . . . three fifths of all other persons."

The idea of counting a slave as three-fifths of a person predated the Constitutional Convention; it arose in response to the problem of taxation under the Articles of Confederation. The

Articles of Confederation provided that taxes would be apportioned among the states on the basis of the value of each state's land. Not surprisingly, states tended to underestimate the value of their land, hindering the ability of the Continental Congress to raise funds.

In February 1783, a special committee of the Continental Congress proposed that taxes be apportioned by population. Southerners did not want to pay taxes on behalf of the approximately 40 percent of the population that was enslaved, so they urged that slaves be counted as one-half or one-fourth of "one freeman." Some northerners argued that, at least for tax purposes, a slave should be counted as the equivalent of a freeman; James Madison proposed a 3 to 4 ratio. Finally, compromise was reached on the 3 to 5 ratio. An amendment to the Articles of Confederation was proposed on April 18, 1783, stating that expenses would be apportioned

> *in proportion to the whole number of white and other free citizens and inhabitants, of every age, sex and condition, including those bound to servitude for a term of years, and three-fifths of all other persons not comprehended in the foregoing description, except Indians not paying taxes, in each State. . . .*

Southern states agreed to the use of three-fifths out of fear that "if an attempt should be made to alter or amend the mode of fixing the quota, those very men would again talk of a Slave being equal to a white man." Although the amendment eventually passed the Continental Congress, it was rejected by New Hampshire and Rhode Island and never became part of the Articles of Confederation. Nevertheless, in 1786 Congress began on its own to use the ratio three-fifths in its apportionment of federal expenses.

The issue resurfaced at the Constitutional Convention, when the framers debated how the population should be calculated for determining representation in the House of Representatives. Although the starting point was the three-fifths figure, there was much dispute. James Wilson of Pennsylvania argued that he "did not see on what principle the admission of blacks in the proportion of three fifths could be explained." He pointed out the absurdity of the ratio: "Are [slaves] admitted as Citizens? then why are they not admitted on an equality with White Citizens? are they admitted as property? then why is not other property admitted into the computation?" Governor Robert Morris objected to including slaves at all in representation calculations, on the grounds that such use would "give such encouragement to the slave trade."

Delegates from slaveholding states threatened to leave the convention if slaves were not included in representation. William Davie of North Carolina said that his state would "never confederate on any terms that did not rate [slaves] at least as three-fifths. If the Eastern States meant therefore to exclude them altogether, the business was at an end."

One way that compromise finally was reached was to tie the population calculation for representation purposes to that used for taxation. In theory, this balanced northern and southern interests. The slaveholding states wanted slaves counted heavily to inflate their population figures for the House of Representatives, but not counted for taxation assessments. Northerners, of course, wanted the opposite. On July 12, the final language—once again settling on the three-fifths compromise—was approved by a vote of 6 to 2, with delegates from Massachusetts and South Carolina divided.

This turned out to be a much less balanced trade-off than was supposed originally: "The principle of representation is constant and uniform; the levy of direct taxes is occasional and rare."

Essentially, almost all pre-Civil War taxation from the federal government was indirect, either tariffs on foreign goods or excise taxes, so the purported extra cost to the South never materialized. Meanwhile, the South enjoyed a numerical edge not only in the House of Representatives but in the electoral college as well.

During ratification debates, some pointed out the circuitous nature of the constitutional language: "What a strange and unnecessary accumulation of words are here used to conceal from the public eye, what might have been expressed in the following concise manner. Representatives are to be proportioned among the states respectively, according to the number of freemen and slaves inhabiting them, counting five slaves for three free men."

The defense of this provision in *The Federalist Papers* is noteworthy for being both uncomfortable and lukewarm:

> *Let the case of the slaves be considered, as it is in truth, a peculiar one. Let the compromising expedient of the Constitution be mutually adopted, which regards them as inhabitants, but as debased by servitude below the equal level of free inhabitants, which regards the SLAVE as divested of two fifths of the MAN.*

By contrast, Justice Joseph Story would later extol the virtues of the plan, as a masterpiece of harmonious reconciliation:

> *Viewed in its proper light, as a real compromise, in a case of conflicting interests, for the common good, the provision is entitled to great praise for its moderation . . . and its tendency to satisfy the people, that the Union, framed by all, ought to be dear to all, by the privileges it confers, as well as the blessings it secures.*

The three-fifths provision hardly could be said to have been "dear to all." Until the Civil War, the provision served to strengthen the power of the slaveholding states. The South's share of the House of Representatives, which would have been 41 percent had only free persons been counted, was more than 46 percent under the three-fifths rule.

As Judge A. Leon Higginbotham Jr. noted, the provision's destructive effects were both symbolic and practical:

> *The three-fifths compromise had given constitutional sanction to the fact that the United States was composed of some persons who were free and others who were not. It set up the principle that a man who lived among slaves had a greater proportion in the election of representatives than the man who did not. It acknowledged slavery and rewarded the slave owners.*

The immediate numerical evil has been undone by the Thirteenth and Fourteenth Amendments. Principles of full citizenship and one-person, one-vote now define our legal landscape.

But the lessons of the three-fifths compromise continue. Not only are we reminded of the stain that slavery inflicted on the framing of the Constitution; we are instructed to consider the many other numerical choices made by the framers. The Constitution still contains numerous provisions that reveal the power of numbers both to communicate philosophy and to further particular goals.

# POLITICAL NUMERACY

# 1

# Logic (Healthy and Ill)

**It reminds me of an answer given some years ago in the School at Oxford, when the Examiner asked for an example of a syllogism. After much patient thought, the candidate handed in:**

> **"All men are dogs;**
> **All dogs are men;**
> ***Therefore,* All men are dogs."**

**This certainly has the form of a syllogism. . . . And it has the great merit . . . that, if you grant the premises, you cannot deny the conclusion. Nevertheless, I feel bound to add that it was *not* commended by the Examiner.**

***—Reverend Charles Dodgson***

The preceding is a lesson on the uses of logic, given by the Reverend Charles Dodgson, who is better known by his pseudonym, Lewis Carroll. The ability to think logically is essential for anyone who wants to understand the Constitution. Teachings from the world of mathematics present both the power of logic and the dangers inherent in utilizing it to confront difficult constitutional issues.

One of the foundations of the mathematical method is that knowledge leads to more knowledge. By the power of logical reasoning, simple truths can lead to countless conclusions of complexity and subtlety. In about 300 B.C., Euclid wrote the *Elements,* which has been called "the most influential textbook of all time." From a mere handful of axioms, Euclid was able to establish hundreds of theorems, encompassing all the important propositions of Greek mathematics. It is not so much the propositions, as the techniques of proof, which make his work of enduring importance.

The critical first step, of course, is the selection of axioms. An axiom can be defined as "a statement used in the premises of arguments and assumed to be true without proof." We must take the correctness of axioms for granted because we have to start somewhere. Some proposition must be the initial one, the one from which the others flow.

Similarly, we never are afforded the luxury of having all our terms defined. Definitions, after all, involve describing one term in reference to another. Unless we are willing to allow circular definitions, where two terms define one another, we must accept undefined terms. Euclid, for example, defined a point as "that which has no parts" but neglected to give a useful definition of *parts*.

Once we have decided what to take for granted, we prove the rest from that. Since undefined terms and unproven assertions are not likely to fill us with confidence, the power of a system is sometimes determined by how few axioms and undefined terms it contains. Hence, Aristotle declared that, "other things being equal, that proof is better which proceeds from the fewer postulates."

Choosing the right axioms is an art. Generally, the axioms should be simple and consistent with one another. They should be logically independent of one another, or else one would be more properly considered a theorem of the other. Finally, axioms "must be fruitful; like carefully selected seeds they must yield a valuable crop. . . ."

The "crop" consists of theorems, which are defined as statements that are "derived from premises rather than assumed." Some theorems may be so obvious they could have been axioms; others may well be surprising and counterintuitive. For one example of the power of this method, see on the next page Euclid's proof that there exists an infinite number of prime numbers.

One of the most common forms of logical analysis involves the use of syllogisms, in which two statements (premises) are linked to form a conclusion. The classic (almost clichéd) syllogism is

1. All men are mortal.
2. Socrates is a man.
3. Therefore, Socrates is mortal.

Similar syllogistic forms can draw conclusions from statements that are less absolute but are sometimes, although not necessarily always, true:

1. Some members of the faculty can't be fired.
2. All people who can't be fired can be independent thinkers.
3. Therefore, some members of the faculty can be independent thinkers.

What is so powerful about this form of logical structure is that it leads to an unquestionable solution: If the premises are all true, then the conclusion must be true as well. Now, not any connection of statements will work. Obviously, if the logical argument is not well formulated, as in the story told by the Reverend Charles Dodgson at the beginning of this chapter, the result is laughable. But if the form is correct, the conclusion is inescapable.

The correctness of the form can be tested by viewing the syllogism in abstract form. The abstract quality of this system per-

### Euclid's Proof That There Are Infinitely Many Prime Numbers

A prime number is a whole number that is only divisible by 1 and itself. Thus, 2, 3, 5, and 7 are all prime, because no numbers divide them exactly except themselves and 1. By contrast, 21 is not prime, because it is divisible by 3 and 7. (By formal definition, 1 is considered neither a prime nor a composite number.)

To prove that there are an unlimited number of primes, Euclid used a technique known as reductio ad absurdum: Assume the opposite of what you want to prove and then prove that a contradiction must inevitably arise.

Euclid's proof begins by assuming that there are only a finite number of primes. With any finite collection of numbers, one must be the largest. We will call it $P$. Next, we can take all of the prime numbers from 2 through $P$, multiply them together, add 1 to that product, and call that new number $Q$ ($Q = 2 \times 3 \times 5 \times 7 \ldots \times P + 1$). We know that there cannot be any prime from 2 to $P$ that divides evenly into $Q$, because each division would leave a remainder of 1.

If $Q$ is prime, we have our contradiction, since $Q$ is a prime larger than $P$. If $Q$ is not prime, it must be divisible by some prime number larger than $P$, since we have already ruled out the prime numbers up to $P$. Either way, we have found some prime number larger than P, either $Q$ or a factor of $Q$.

Therefore, we have contradicted our premise that there is a largest prime number. All that is left to do is to conclude the opposite: There is no largest prime. That means that the primes are infinite. At this point, we can proudly declare that we are finished by announcing Q.E.D. (*quod erat demonstrandum*, "that which was to be proved").

mits the substitution of symbols for the words in each syllogism, which leads to their universal acceptability.

For example, the syllogism about Socrates can be restated as

1. Every *M* is a *P*.
2. *S* is an *M*.
3. Therefore, *S* is a *P*.

The independent-minded faculty syllogism would be restated as

1. Some *M* are *P*.
2. Every *P* is an *S*.
3. Therefore, some *M* are *S*.

It does not matter what the *M, P*, or *S* represents. Each of these syllogisms is correct. This abstraction not only gives logic its strength, it also reveals its ultimate weakness. In the words of one of the leading logicians of the twentieth century, Bertrand Russell,

> *Pure mathematics consists entirely of such asservations as that, if such and such a proposition is true of* anything, *then such and such another proposition is true of that thing. It is essential not to discuss whether the first proposition is really true, and not to mention what the anything is of which it is supposed to be true. . . . If our hypothesis is about* anything *and not about some one or more particular things, then our deductions constitute mathematics. Thus,* mathematics may be defined as the subject in which we never know what we are talking about, nor whether what we are saying is true.

Both for mathematicians and for others who use logical reasoning, the danger is glaring: if your initial axioms are incorrect, then your conclusions are not guaranteed. You may be arguing perfectly logically but still end up with a ridiculous conclusion because of an initial poor choice of axioms. Thus, some have remarked that "Logic is the art of going wrong with confidence."

÷ ÷ ÷

Without question, the most successful application of logical analysis to the axioms of political science is the Declaration of Independence, that "famous 'mathematical' document." The Declaration of Independence is a paradigm of deductive reasoning, where political axioms are announced, then specific facts that show the applicability of those axioms are described, and, finally, the conclusion of independence is proclaimed.

In Thomas Jefferson's library was a book by a John Harris, entitled *Lexicom Technicum,* which defined *axiom* as "such a common, plain, *self-evident* and received Notion, that it cannot be made more plain and evident by demonstration." The ringing phrase "We hold these truths to be self-evident" is really a statement that the "truths" are to be considered the political equivalents of Euclid's axioms.

There were four such "truths" in the Declaration:

1. "[T]hat all men are created equal"
2. "[T]hat they are endowed by their Creator with certain inalienable rights; that among these, are life, liberty, and the pursuit of happiness"
3. "that, to secure these rights, governments are instituted among men, deriving their just powers from the consent of the governed"
4. "[T]hat, whenever any form of government becomes destructive of these ends, it is the right of the people to alter or to abolish it, and to institute a new government"

There is no need to supply any proof for these "truths." They are the "Elements" of our political system.

Jefferson then presented a long list of facts, demonstrating that the first two axioms had been violated by "a long train of abuses and usurpations. . . ." According to the third axiom, governments are instituted to prevent such violations. According to the last axiom, when such violations occur, the people have the right to institute a new government. "We," the Declaration concludes, "therefore, solemnly publish and declare, that these united colonies, are, and of right ought to be, free and independent states." Q.E.D.

+ + +

In his defense of the Constitution, Alexander Hamilton also relied on the power of axiomatic analysis. In Federalist 31, Hamilton explained that "In Disquisitions of every kind, there are certain primary truths, or first principles, upon which all subsequent reasonings must depend." He stated that this was true not only of "maxims in geometry," but with "these other maxims in ethics and politics . . . [such as] the means ought to be proportioned to the end [and] that every power ought to be commensurate with its object. . . ."

For example, he argued that because of its responsibility for guaranteeing national defense and "securing the public peace against foreign or domestic violence," the federal government needed the unqualified ability to raise money through taxation so as to perform its tasks. Hamilton criticized the opponents of the Constitution who denied the logical conclusions that flowed from the basic political principles: "The obscurity is much oftener in the passions and prejudices of the reasoner than in the subject."

Hamilton acknowledged that "it cannot be pretended that the

principles of moral and political knowledge have, in general, the same degree of certainty with those of the mathematics." Nevertheless, he maintained that the major limitation on the use of political logic was the greater difficulty in which the results of its syllogisms were accepted than in the ethereal world of mathematics: "The objects of geometrical inquiry are so entirely abstracted from those pursuits which stir up and put in motion the unruly passions of the human heart, that mankind, without difficulty, adopt not only the more simple theorems of the science, but even those abstruse paradoxes which . . . are at variance with the natural conceptions which the mind . . . would be led to entertain upon the subject."

## Constitutional Axioms

*[W]e must never forget, that it is a* constitution *we are expounding.*

—CHIEF JUSTICE JOHN MARSHALL,
*McCulloch v. Maryland* (1819)

Why is the Constitution so short? Because in many ways the Constitution provides only the axioms of our system of government, which we then are required to interpret and explain. According to Chief Justice William Rehnquist, in interpreting the Constitution, the Supreme Court tries to "discern among its 'essential postulates,' a principle that controls the present cases." This concept is captured in John Marshall's famous explanation of why interpreting the Constitution is a different venture than other forms of legal interpretation. Marshall stated that any document that contained all of the details of a government's structure and power would "partake of the prolixity of a legal code, and could scarcely be embraced by the human mind." Thus, by its very nature, a constitution requires that only the "great outlines should be marked, important objects designated," so that the minor ingredients "which compose those objects be deduced from the nature of the objects themselves."

Many of the great opinions penned by Marshall were styled deliberately as Euclidean proofs. In *McCulloch v. Maryland,* Marshall laid out his argument that states lacked power to tax a federally incorporated bank, as follows:

1. "This great principle is that the constitution and the laws made in pursuance thereof are supreme . . . and cannot be controlled by [the states]."
2. "From this, which may almost be termed an axiom, other propositions are deduced as corollaries . . . that a power to destroy, if wielded by a different hand, is hostile to and incompatible with these powers to create and to preserve."
3. "That the power of taxing it by the States may be exercised so as to destroy it, is too obvious to be denied."

In *Gibbons v. Ogden,* Marshall essentially apologized for his extended "proof" that federal law preempted a state-authorized monopoly to run a ferry in New York waters:

> *The Court is aware that, in stating the train of reasoning by which we have been conducted to this result, much time has been consumed in the attempt to demonstrate propositions which might have been thought axioms. . . . The conclusion to which we have come depends on a chain of principles which it was necessary to preserve unbroken; and, although some of them were thought nearly self-evident, the magnitude of the question, the weight of character belonging to those from whose judgment we dissent, and the argument at the bar, demanded that we should assume nothing.*

≠ ≠ ≠

One of the greatest examples of the use of logic to reach a conclusion of constitutional interpretation is Oliver Wendell Holmes's defense of "the marketplace of ideas." His oft-quoted analysis includes not only two competing syllogisms but reference to a mathematical proof as well:

> *Persecution for the expression of opinions seems to me perfectly logical. If you have no doubt of your premises or your power and want a certain result with all your heart you naturally express your wishes in law and sweep away all opposition. To allow opposition by speech seems to indicate that you think the speech impotent, as when a man says that he has squared the circle, or that you do not care whole-heartedly for the result, or that you doubt either your power or your premises. But when men have realized that time has upset many fighting faiths, they may come to believe even more than they believe the very foundations of their own conduct that the ultimate good desired is better reached by free trade in ideas—that the best test of truth is the power of the thought to get itself accepted in the competition of the market, and that truth is the only ground upon which their wishes safely can be carried out. That at any rate is the theory of our Constitution. It is an experiment, as all life is an experiment.*

As with Euclid's proof of the infinitude of primes, Holmes begins by presuming the opposite of what he intends to prove. He states that censorship seems to be logical and that one "naturally" would want to sweep away opposition. That certainly would be the case, he says, if the only reasons for allowing opposition were that (1) the opposing speech was "impotent," that is, unable to prevail; (2) the speech was on an unimportant topic; (3) those in power doubted their ability to enforce a ban; or (4) those in power lacked faith in their own position.

But, Holmes says, history has shown that the ideas about which people were completely confident often turned out to be incorrect. (Note that Holmes uses volumes of history in his page of logic.) Holmes hardly needed to reference the calamitous event of his lifetime, the Civil War, in which so many were willing to die over their "faith" in the correctness of slavery. Therefore, he says, it is logical to believe in the premise that the free sharing of ideas will lead to the "truth" rather than in the competing premise that the "truth" is possessed by any one individual, even oneself. And this, he adds, is the "theory" of the Constitution.

Holmes concludes his analysis by highlighting the significance of using political, rather than mathematical, axioms. There never can be the same certainty in politics as there is in mathematics. The effectiveness of the marketplace of ideas is not a guarantee, he says, but merely "an experiment."

× × ×

Holmes's example of speech that is "impotent," was drawn from an important, and for him relatively recent, mathematical proof. Squaring the circle represented one of the oldest challenges in math. For over 2,000 years, people had been trying unsuccessfully to solve a problem posed by the ancient Greeks: Given a circle, construct a square with the same area, using only a straight edge (an unmarked ruler) and a compass. Meanwhile, many other constructions had been created, such as bisecting angles and inscribing a regular hexagon in a circle. But squaring the circle baffled the greatest mathematical minds of two millennia.

Finally, in 1882, Ferdinand Lindemann, a German mathematician, proved that such a construction was impossible. He began his proof by observing that only certain lengths can be constructed with the straight edge and compass. Specifically, the only permissible lengths are those which can be computed using integers (whole numbers) and addition, subtraction, multiplication, division, and square roots. Numbers so computed are known as algebraic numbers. Numbers that are not algebraic are called transcendental numbers.

Lindemann then established that $\pi$ was transcendental. This led to the final proof. If a circle has a radius of 1, its area ($\pi r^2$) would be $\pi$. Thus, to construct a square with the same area ($\pi$) would require creating a side whose length was the square root of that area, $\sqrt{\pi}$. Because $\pi$ is transcendental, $\sqrt{\pi}$ is transcendental. Because the required length is not an algebraic number, it cannot be constructed using a straight edge and compass. Thus, the desired square cannot be constructed. So, "when a man says that he has squared the circle," his speech is impotent because not only is he incorrect but none of his protestations to the contrary can ever alter that reality.

= = =

In the words of Oliver Wendell Holmes Jr., "The life of the law has not been logic: it has been experience." In practice, he never was one to discard the use of logic. He merely was emphasizing that law is not primarily the creature of axiomatic development, but also incorporates "[t]he felt necessities of the time, the prevalent moral and political theories, intuitions of public policy, avowed or unconscious, even the prejudices which judges share with their fellow-men. . . ." As Judge Richard Posner has noted, although "it plays a role, a *critical* role . . . [l]ogic will not decide the most difficult cases."

Two infamous Supreme Court cases reveal the perils of the reliance on logic alone. In *Dred Scott v. Sanford,* the Court ruled that African Americans, whether slave or free, never could be citizens of the United States. In one of the most chilling paragraphs in U.S. law, Chief Justice Taney explained the premise for their constitutional plight:

> *They had for more than a century before been regarded as beings of an inferior order, and altogether unfit to associate with the white race, either in social or political relations; and so far inferior, that they had no rights which the white man was bound to respect. . . . This opinion was at the time fixed and universal in the civilized portion of the white race. It was regarded as an* axiom *in morals as well as in politics, which no one thought of disputing, or supposed to be open to dispute. . . .*

The "axiom" of racial inferiority, accepted by the Court, led to what is widely considered "one of the great disasters in the history of the Supreme Court." One lesson from *Dred Scott* is that if you start with an incorrect axiom, you are unable to reason intelligently. Such an analysis has been likened to a house of cards, built instead from a foundation of axioms, with theorem piled on top of theorem: "If one element were faulty, then the whole structure could come tumbling down."

But logical analysis also can help reveal unspoken prejudices. In *Korematsu v. U.S.,* the Court upheld the wartime military order which led to the evacuation of Japanese Americans from their homes on the West Coast and forced them into relocation centers. The Court accepted the judgment of military authorities that:

1. "[T]here were disloyal members of that [Japanese] population whose number and strength could not be precisely and quickly ascertained."
2. "[S]uch persons could not be readily isolated and separately dealt with, and constituted a menace to the national defense and safety."

This can be thought of as two different syllogisms, with the "theorem" that is proved in the first syllogism used in the next:

*Syllogism A*

1. It is necessary to isolate all disloyal persons during time of war.
2. Some Japanese Americans are disloyal.
3. Therefore, it is necessary to isolate some Japanese Americans during time of war.

*Syllogism B*

1. It is necessary to isolate some Japanese Americans during time of war.
2. The military cannot separate loyal Japanese Americans from disloyal Japanese Americans.
3. Therefore, it is necessary to isolate all Japanese Americans during time of war.

Syllogism A is logically sound, but that cannot be said for syllogism B. The problem comes from its second (or minor) premise, the inability to separate loyal from disloyal Japanese Americans. The Court did not require proof of this point but accepted the bald assertion of the military commander, Commanding General J. L. DeWitt. In fact, according to internal Justice Department documents released after the war, General DeWitt's report on the necessity of wholesale exclusions of Japanese Americans, was filled with "intentional falsehoods," that were "highly unfair to this racial minority."

One way to reveal the weakness in the *Korematsu* decision is to expose the inconsistency within the Court's own logic. First, as Justice Frank Murphy argued in dissent, the problem of mixed loyalties affected American descendants from each of our World War II opponents: "Similar disloyal activities have been engaged in by many persons of German, Italian and even more pioneer stock in our country."

If the opinion were not infected with racial hostility, we should be able to substitute comparable nationalities in the syllogisms. Thus, according to syllogism A, it would have been necessary to isolate some German Americans and some Italian Americans. Lacking even a shred of evidence that it would have been any easier or quicker to separate the loyal from the disloyal within these two groups, under the logic of *Korematsu,* there should have been wholesale evacuations of these groups as well.

The failure to apply the same logic to other groups indicates that racial animus was in fact pollluting the legal analysis. As Justice Murphy explained,

> *The main reasons relied upon by those responsible for the forced evacuation . . . [appear] to be largely an accumulation of much of the misinformation, half-truths and insinuations that for years have been directed against Japanese Americans by people with racial and economic prejudices—the same people who have been among the foremost advocates of the evacuation.*

Logic is a marvelous tool. But if it is based on incorrect assumptions, as in *Dred Scott,* or if it is done badly, as in *Korematsu,* the results can be flawed. However, unlike a flawed geometric proof, a flawed constitutional analysis can have disastrous effects on human beings.

## Burdens of Proof

> *It is better that ten guilty persons escape than one innocent suffer.*
>
> —Sir William Blackstone,
> *The Law of England* (1807)

When a mathematical proof is done correctly, there is no questioning the result. There *are* an infinite number of primes.

You *cannot* square a circle. I am certain. In law, however, there is no such comfort. During a trial, even after a lawyer has proved a case, uncertainty abounds.

Mathematics also recognizes uncertainty, which is measured by probability. If an event is certain to occur, it is given a value of 1, and if it is impossible, the value is 0. A value of .5 (or 50 percent), indicates that an event is as likely to occur as not. The greater the likelihood, the higher the value, up to 1. The less likely the outcome, the lower the value, down to 0.

Thus, probability is used to make inferences, not guarantees. If I flip a coin, I don't know whether to expect heads or tails. If I roll two dice, I will be surprised, though not shocked, to get snake eyes (two 1s). If I am playing poker, I would be amazed to be dealt four aces, and even more astonished if simultaneously you had been dealt a straight flush.

When we make assumptions, therefore, we are in some sense calculating probabilities. But this is not the case, however, when we discuss the "presumption of innocence." We do not mean that a particular defendant is more likely innocent than guilty. In fact, we could reach the opposite conclusion, and, in the words of Supreme Court Justice John Paul Stevens, make the "empirical judgment that most persons formally accused of criminal conduct are probably guilty." Although we say that a criminal defendant is "innocent until proven guilty," pretrial detainees can be confined in prisons and subject to substantial restraints on their freedom.

What the presumption of innocence does describe is the burden that is placed on the government to prove its case before the defendant can be convicted of a crime. The Constitution mandates that no person can be convicted of a crime except on proof "beyond a reasonable doubt." If the prosecution does not establish every fact necessary to constitute the crime beyond a reasonable doubt, the defendant is acquitted.

In a civil trial, the standard is much lower. To prevail, all one party has to do is establish the necessary facts "by a preponderance of the evidence." This is generally understood to mean that the fact finder (jury or judge) has decided that the claim of one side is more likely true than not. Thus, it is logically—and legally—consistent for O. J. Simpson to be found both "not guilty of" and "responsible for" the killing of Nicole Simpson and Ron Goldman.

If the purpose of any trial, criminal or civil, is to discover the truth, then why is there a different standard for the two types of cases? The answer lies not in the desire for truth but in the inevitability of error. The Supreme Court has recognized that even the "beyond a reasonable doubt standard" is inherently "probabilistic . . . [since] the fact finder cannot acquire unassailably accurate knowledge of what happened. Instead all the fact finder can acquire is a belief of what *probably* happened." Knowing that mistakes will happen, the issue is not simply how do we reduce mistakes but which specific kind of mistakes do we most want to avoid.

Suppose, for example, that there was a medical test to determine whether patients had a particular disease, and researchers believed that a higher test score correlated to an increased likelihood of having the disease. There are two situations where the test could be wrong. First, healthy patients could be diagnosed with the disease. In statistics, this would be termed a Type I error. Alternatively, diseased patients could be diagnosed as healthy, which would be termed a Type II error.

| **Diagnosed Condition** | **Actual Condition** | |
|---|---|---|
| | *Healthy* | *Diseased* |
| *Healthy* ➤ | Correct | Type II error |
| *Diseased* ➤ | Type I error | Correct |

As patients have a range of scores on this test, a cutoff point is needed so that we can say that above *this* score, the patient will be diagnosed with the condition. Let us say that a particular cutoff point is chosen, such that patients with a score of 60 or greater will be deemed to have the disease. A table indicating how well the test worked might look like this:

| **Diagnosed Condition** | | **Actual Condition** | |
|---|---|---|---|
| | *Healthy* | *Diseased* | *Total* |
| *Healthy* ➤ | 350 | 10 | 360 |
| *Diseased* ➤ | 40 | 200 | 240 |
| *Total* ➤ | 390 | 210 | 600 |

The Type I error rate is the total of healthy patients incorrectly diagnosed divided by the total healthy population. For the above example, the false positive rate would be 40/390, or 10.3 percent. The type II error rate, the total of incorrectly diagnosed diseased patients divided by the total diseased population, is 10/210, or 4.8 percent.

If it is determined that one kind of error is worse than the other, the more harmful type of error can be reduced by adjusting the cutoff point.

The following three graphs show the effect of changing the cutoff. In all three graphs, we will use two curves. The curve beginning at the left shows the varying test scores of healthy people, and the curve to the right shows the scores of those with the disease. (In reality, the ends of the curves never actually touch the bottom of the graph; there is always some possibility of error.) Note that no matter which cutoff score you choose, you will make some errors.

In graph 1, the cutoff line was chosen to equalize the rates of

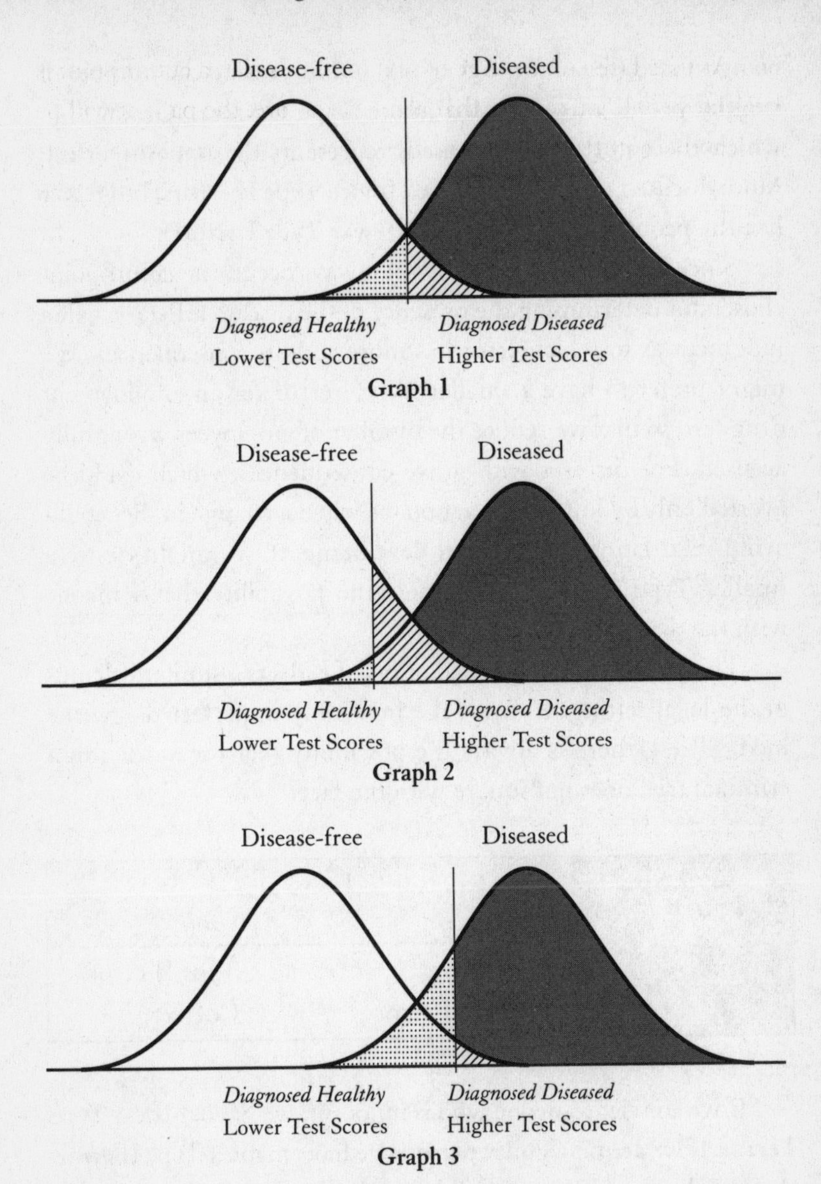


Type I and Type II errors. The slashed area represents Type I errors, and the dotted area represents Type II errors. In graph 2 the cutoff point is lowered, resulting in fewer diseased persons being missed (fewer Type II errors) but increasing the number of healthy people misdiagnosed (more Type I errors). Graph 3, in which the cutoff point is raised, represents the opposite effect: More diseased persons are missed (more Type II errors) but fewer healthy people are misdiagnosed (fewer Type I errors).

Since both kinds of errors will always occur, the cutoff point chosen for determining the presence of the disease reflects a value judgment as to which error has more serious consequences. We might prefer to have a smaller Type I error for an employment drug test, so that we reduce the number of employees wrongfully accused. For diseases with grave consequences, which could be averted only by immediate action (as when a change in diet could avoid retardation during fetal development), we might desire a smaller Type II rate, to minimize the possibility that someone with the disease goes undiagnosed.

The choice of the standard of proof reflects a similar calculus in the legal setting. As with the inevitably imperfect diagnostic medical test, there is always the possibility that the verdict in a criminal trial does not square with the facts.

| **Verdict** | **Facts** | |
|---|---|---|
| | *Innocent* | *Guilty* |
| *Not Guilty* ➤ | Correct | Type II error |
| *Guilty* ➤ | Type I error | Correct |

If we convict someone who is innocent, we have made a Type I error. If we acquit a guilty person, we have made a Type II error. And we know that errors will be made.

Adjusting the standard of proof affects the frequency of each type of error. Just as raising the cutoff point resulted in fewer healthy people being diagnosed as diseased, the higher we make the standard of proof, the fewer the innocent people who will be found guilty. The cost, of course, is that then more guilty people will be acquitted.

≥ ≥ ≥

For criminal cases, the Constitution requires the highest standard, beyond a reasonable doubt, because of "a fundamental value determination of our society that it is far worse to convict an innocent man than to let a guilty man go free." This principle predates the Constitution, as reflected in Sir William Blackstone's admonition that English law recognized that it was preferable for ten guilty persons to escape than for one innocent person to be convicted wrongfully. The Supreme Court has explained that this balance reflects the fact that the accused has a far greater stake in a criminal trial than even the government: "Where one party has at stake an interest of transcending value—as a criminal defendant his liberty—[the] margin of error is reduced as to him by the process of placing on the other party the burden of persuading the fact finder . . . of his guilt beyond a reasonable doubt."

By contrast, in a civil suit between two parties, where the plaintiff alleges that the defendant is responsible for some monetary loss, there is no reason to believe that either plaintiff or defendant generally suffers the greater harm from an erroneous decision.

| Verdict | Facts | |
|---|---|---|
| | *Not Responsible* | *Responsible* |
| *Not Responsible* ➤ | Correct | Type II error |
| *Responsible* ➤ | Type I error | Correct |

Unlike in a criminal trial, an incorrect finding that the defendant was responsible would produce no obviously greater hardship than an incorrect finding of "not responsible." When "the social disutility of error in either direction is roughly equal," the risks of an erroneous determination are equalized by the preponderance of the evidence standard.

The diverging standards explain the following Supreme Court decision which, at first, may seem contradictory. A defendant was convicted of one count of selling cocaine but acquitted on charges involving a second sale. Nonetheless, the Supreme Court permitted the judge to impose a harsh sentence based on the defendant's participation in the second as well as the first transaction. As the Supreme Court noted, the acquittal did not prove that the defendant was "innocent"; it merely signified that the government had not met its elevated burden of proof. Unlike convictions, however, sentencing decisions can include many factors, either mitigating or aggravating, that are not established with that high level of certainty. It is not contradictory to believe that something is more probable than not, yet still have reasonable doubt.

≠ ≠ ≠

The Supreme Court has even created a third standard, "clear and convincing evidence," which it has described as "a middle level of burden of proof." This standard is greater than a mere preponderance of the evidence but is not as demanding as beyond a reasonable doubt. The Court has required the government to prove its case by clear and convincing evidence, in several noncriminal cases where government action threatened a significant deprivation of liberty, such as civil commitment, deportation, and denaturalization. This increased burden is chosen in order that a larger part of the risk of error be imposed on the government: "The individual should not be asked to share equally with society the risk of error when the possible injury to the individual is significantly greater than any possible harm to the state."

The question of when an injury is "significantly greater" is not always self-evident. For example, in *Santosky v. Kramer,* the Court struggled with the wrenching issuing of adjudicating the loss of parental rights. In that case, three children were removed from their parents' custody after the local department of social services found evidence of abuse, malnutrition, and neglect. The issue for the Supreme Court was to determine the standard of proof the government needed to establish before permanently terminating parental rights. A majority of the Court found that "a natural parent's desire for and right to the companionship, care, custody, and management of his or her children is an interest far more precious than any property right." Thus, the Court concluded, because the harm from permanent deprivation far outweighed any competing government interest, the "clear and convincing" standard must be reached before the termination of parental rights.

In dissent, then-Associate Justice William Rehnquist argued that the harm from erroneously maintaining parental rights was being underestimated: "If the Family Court makes an incorrect factual determination resulting in a failure to terminate a parent-child relationship which rightfully should be ended, the child involved must return either to an abusive home or to the often unstable world of foster care." Therefore, he stated, the two types of errors should be viewed as having equal seriousness, and a "preponderance of the evidence" standard should have been utilized to determine what was best for the children.

A second case, *Cruzan v. Director, Missouri Department of Health,* dealt with parents who wished to terminate the life support system of their comatose daughter. The Supreme Court permitted the State of Missouri to overrule the desires of the parents unless they could prove "by clear and convincing evidence" that their daughter would have wanted to avoid further medical treatment. The Court emphasized the state's great interest in the "protection and preservation of human life" and concluded that "An erroneous decision to withdraw life-sustaining treatment . . . is not susceptible of correction. . . . An erroneous decision not to terminate [could be corrected by] the possibility of subsequent developments such as advancements in medical science, [or] the discovery of new evidence regarding the patient's intent. . . ." This time, Justice William Brennan dissented, contending that comparably serious harm would follow from both kinds of error: "An erroneous decision not to terminate life support . . . robs a patient of the very qualities protected by the right to avoid unwanted medical treatment. His own degraded existence is perpetuated; his family's suffering is protracted; the memory he leaves behind becomes more and more distorted."

In both cases, the "clear and convincing" standard was chosen by the Court over the "preponderance of evidence" standard in order to reflect a determination that greater harm resulted from one type of error than another. While mathematical analysis can tell the Court which standard will best implement such a determination, the value-laden evaluation of the comparative seriousness of the harms involved is emphatically not a mathematical decision.

+ + +

Despite its usefulness elsewhere, mathematics is remarkably ineffective at quantifying phrases such as "beyond a reasonable doubt" or "clear and convincing evidence." Some have tried to develop numerical equivalents for differing legal standards, suggesting, for example, a finding that the "reasonable doubt" standard was met would mean that the fact finder determined that guilt was at least 95 percent probable. Similarly, a finding of "clear and convincing evidence" would mean a determination that a result was reached with at least 70 percent certainty, while "preponderance of the evidence" would mean with at least 50 percent. Even if one were to assume that the average juror were comfortable with the application of abstract mathematical concepts, the reality is that the strength of belief is one of the many things in life that cannot be quantified. As Professor J. H. Wigmore wrote, "No one has yet invented or discovered a mode of measurement for the intensity of human belief."

If a jury heard conflicting testimony between two witnesses, the arresting officer and the defendant, it would be useless to demand a decision based on "95 percent certainty." On the other hand, the following nonarithmetic instruction by a judge could indeed enable a juror to reach the constitutionally appropriate decision:

> *Proof beyond a reasonable doubt is proof that leaves you firmly convinced of the defendant's guilt. There are very few things in this world that we know with absolute certainty, and in criminal cases the law does not require proof that overcomes every possible doubt. If, based on your consideration of the evidence, you are firmly convinced that the defendant is guilty of the crime charged, you must find him guilty. If on the other hand, you think there is a real possibility that he is not guilty, you must give him the benefit of the doubt and find him not guilty.*

This book is premised on the belief that there are many legal ideas that can be explained or clarified by mathematics. But great care must be taken that we never lose sight of the essential but unquantifiable factors that should be part of our most important decisions. Mathematical reasoning should not be confused with reducing intrinsically complex phenomena to an artificial arithmetic quantity.

# 2 Majority Rules

**The search of the great minds of recorded history for the perfect democracy, it turns out, is the search for a chimera, for a logical self-contradiction.**

***—Paul Samuelson,***
***quoted in* Archimedes' Revenge,**
***Paul Hoffman (1988).***

The Constitution, in the first three words of its Preamble, declares that the fundamental source of its authority is derived from "We the People." The American ideal of "government of the people, by the people, and for the people," runs through our nation's history and constitutional jurisprudence. There is no more basic principle of our governmental system than that "the fabric of American empire ought to rest on the solid basis of THE CONSENT OF THE PEOPLE. The streams of national power ought to flow immediately from that pure, original fountain of all legitimate authority."

"The consent of the people" cannot be understood to be the same as the consent of an individual person. Each individual either consents or does not, but within a group, disagreements will occur. According to Thomas Jefferson, the will of the group is, by natural right, to be determined by majority rule: "Individuals exercise [the right of self-government] by their single will: collections of men by that of their majority; for the law of the *majority* is the natural law of every society of men."

The spirit of democracy often is described as a belief in majority rule. For example, Alexander Hamilton wrote in Federalist 22, "[T]he fundamental maxim of republican government . . . requires that the sense of the majority should prevail." And the Supreme Court wrote with peculiar understatement, "Logically, in a society ostensibly grounded on representative government, it would seem reasonable that a majority of the people of a State could elect a majority of that State's legislators."

The reality is that in our constitutional system majority rule is not necessarily the rule. Though it might come as a surprise to many Americans, the Supreme Court has held, "There is nothing in the language of the Constitution, our history, or our cases that requires a majority always prevail on every issue." And, if one wanted to be precise, there are *no* decisions under our Constitution that must be decided by a majority of citizens. This is, in fact, an inevitable consequence of representative government and of our system of federalism.

Imagine that a state, such as New Mexico or West Virginia, is electing three members to the House of Representatives. Under Article I, Section 2, each congressional district must be the same size. It certainly is possible that there are more voters of one party, say Democrats, in the state but that the voters are not evenly distributed. Thus, if Democrats dominate one district by a wide margin, but Republicans hold a narrow edge in the other two, a majority of the state's representatives will be Republican, despite a statewide majority of Democrats.

Only if a state, through deliberate gerrymandering, were able to draw its district lines so as to ensure this numerical result in an effective and continuing manner, would the Supreme Court find the "continued frustration of the will of the majority of voters" to be unconstitutional.

Usually, this "frustration" occurs naturally whenever the degree of political dissension is not uniform. Whenever opinions between or within states are divided to varying degrees, the will of the majority of voters may be frustrated. This defeat for the majority of voters is not unconstitutional when it is the result of the inevitable normal workings of a representative system.

≠ ≠ ≠

The framers also realized that even arithmetically pure representation, the idealized situation where representatives accurately reflect the will of the majority, poses serious problems. Most particularly, the very concept of democracy, which prevents kings, dictators, and small groups from imposing their will, not only permits the majority of citizens to impose its will but permits that majority to inflict harm on others as well. As James Madison wrote to Thomas Jefferson,

> *Wherever the real power in a Government lies, there is the danger of oppression. In our Governments the real power lies in the majority of the Community, and the invasion of private rights is chiefly to be apprehended, not from acts of Government contrary to the sense of its constituents, but from acts in which the Government is the mere instrument of the major number of the constituents.*

The danger here is not so much that a single harmful decision will be rendered, because mistakes can be corrected by the next election. The real concern is that the ongoing decision-making process might perpetuate, rather than ameliorate, an invasion of private rights. For example, in a group of 11, if 6 are strongly united, they can dominate the other group members without fear of retribution. If, on the other hand, the membership of the majority is constantly shifting, with each of those 11 periodically in the majority, the risk of domination decreases. Each temporary majority has less time to inflict harm, and the next majority can undo, in large or small measure, an inflicted harm. More importantly, if governing majorities are fluid, a majority of voters will experience minority status, and perhaps a majority may then see the virtue in respecting minority interests. Thus, the key to avoiding intrasocietal oppression is the prevention of a permanent majority interest group, for when "a majority be united by a common interest, the rights of the minority will be insecure."

According to Madison, the first line of defense against a "united" majority is to "extend the sphere" over which the majority must be drawn. He argued that the United States needed a powerful national government, because significantly increasing the size of the electorate would make it much more difficult for any one group to obtain and maintain a working majority: "[Y]ou take in a greater variety of parties and interests [and] you make it less probable that a majority of the whole will have a common motive to invade the rights of other citizens. . . ."

× × ×

When the majority rules, numbers matter, and size is a virtue. The larger the population of the electorate, the more varied the interests are likely to be. The more varied the interests, the harder it will be to have the same group maintain agreement on a wide range of issues for a long period of time. Paradoxically, while your individual voice is less powerful in a large population, your vote may be more powerful.

There is a mathematical measure of political power that determines the frequency in which a voter is a "pivotal player," that is, someone who can convert a losing coalition into an electorally successful one. The so-called Shapley-Shubik index determines the "fraction of power" an individual has, by calculating the percentage of all possible permutations of voters for which the vote of that individual would be pivotal. Intuitively, one would assume that the smaller the group, the greater the frequency in which any one individual would be pivotal.

The important insight of the framers was that this intuition is in fact incorrect. While a small group may have fewer members, it is also more homogeneous and has fewer competing points of view. Thus, it is dangerously easy for a small number of individuals to "concert and execute their plans of oppression." Once that situation occurs, of course, the individuals locked out of that small favored group never can be pivotal. Unable to create a new winning majority, they will have no fraction of power at all.

By contrast, in a larger society, solidified majorities are not only more difficult to obtain but much more difficult to maintain, "by comprehending in the society so many separate descriptions of citizens as will render an unjust combination of a majority of the whole very improbable, if not impracticable." In such a setting, therefore, all individuals and groups of individuals will have, at least occasionally, the opportunity to be the pivotal player of a winning coalition. Their fraction of the total vote may be small, but at least they possess some fraction of the political power.

Such a situation was seen in the civil rights struggles of the early 1960s. African Americans constituted a larger proportion of the population of many southern states than of the United States at large, yet they were disenfranchised and politically oppressed within the relatively small confines of the individual southern states. On a national level, however, with a far larger, more heterogenous population, legislation such as the Civil Rights Act of 1964 and the Voting Rights Act of 1965 was enacted.

A different majoritarian dilemma proved to be one of the most contentious issues at the Constitutional Convention. In a representative democracy, if voters are not evenly distributed, some voters may end up without any real political power. Those at the Convention were faced with 13 states having widely varying populations. Consider just four states, Virginia, Massachusetts, Connecticut, and Rhode Island, and, for simplicity, imagine that Virginia has 100 voters, Massachusetts has 80, Connecticut has 60, and Rhode Island has 10. If representation is on a 10 to 1 basis, then Virginia has 10 votes in the legislature, Massachusetts has 8, Connecticut has 6, and Rhode Island has but 1.

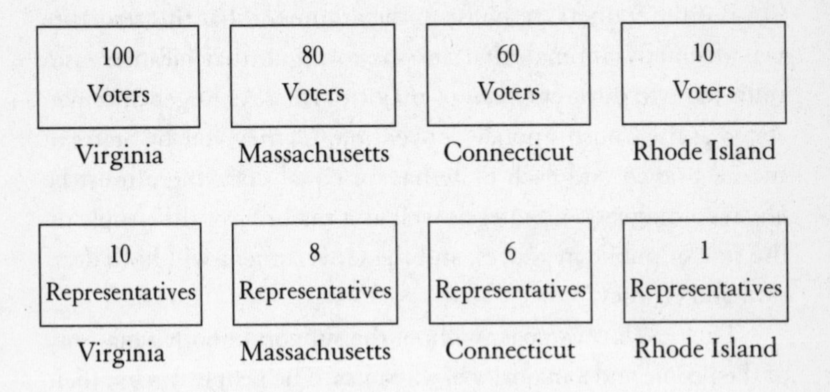


If each state delegation is united behind a common interest and votes as a bloc, the vote of Rhode Island's lone representative would rarely, if ever, be pivotal. The interests of the residents of that tiny state safely could be ignored by the others.

The remedy adopted at the Constitutional Convention, known as the Connecticut Plan, created two distinct representative bodies, one divided by population, the other equally divided by state. The smaller states, then, have the opportunity to be pivotal players in at least one of the two houses.

Since the states have widely varying populations, this bicameralism frequently will thwart the will of the majority. A small fraction of the population, located in the smallest states, can prevent legislation from being passed, even if the proposal is supported by the great majority of Americans. For example, at the time of the first census, the eight smallest states, which controlled 16 of the 30 Senate seats, had less than 25 percent of the nation's population. Thus, less than one-fourth of the population could stymie the will of the rest. The situation is even more extreme in modern times. As of the 2000 census, the 26 smallest states, controlling 52 of the 100 Senate seats, have less than 18 percent of the national population.

But the framers' response to the argument that this structure was distinctly antimajoritarian was to argue that bicameralism reflected two different *kinds* of majority rule. As Roger Sherman stated at the Constitutional Convention, "If they vote by States in the 2d branch and each State has an equal vote, there must be always a majority of States as well as a majority of the people on the side of public measures, and the Government will have decision and efficacy."

Thus, no law can pass without the support of both a majority of the people and a majority of the states. The people can use their power in the House of Representatives to stop small factions from inflicting harm on the majority. The Senate is constructed to prevent the majority who populate the largest states from injuring their fellow citizens.

## The Mathematical Virtues of the Electoral College

Nowhere has the structural limitation on the power of the majority been made more obvious than in the presidential election of 2000. George W. Bush received more than 150,000 fewer popular votes than Al Gore but prevailed by a vote of 271 to 267 in the electoral college. Amid the disputes over hanging chads and uncounted ballots, the public almost lost sight of the fact that for the first time in more than 100 years, the presidency was won by a candidate who garnered fewer popular votes than his opponent. This failure of majority will, however, is the deliberate result of the decisions made by the framers of the Constitution to place the power to select the president in the hands of the electoral college.

The electoral college was created in large part because the delegates to the Constitutional Convention could not agree on any other way to select a president. As James Madison noted after many days of frustrating stalemate, "There are objections against every model that has been, or perhaps can be proposed." Proposals that the presidential selection be made by Congress, state legislatures, or governors had all been rejected on the grounds that the president would be subservient to any political group that appointed him.

Direct popular election was rejected primarily on two unsavory grounds. First, many of the framers possessed a strong antidemocratic distrust of the people. As Elbridge Gerry of Massachusetts argued, "The people are uninformed, and would be misled by a few designing men."

The second complaint about direct popular voting for the president reflects the poison that slavery infused into so much of the Constitutional Convention. Southerners opposed direct election. They feared being outvoted, since the slaves who made up so much of their population were barred from voting. As James Madison said, "The right of suffrage was much more diffusive in the Northern than the Southern States; and the latter could have no influence in the election on the score of the Negroes." There were also fears from smaller states that their choice for president would be outvoted easily by the voters of the larger states.

The use of electors to select the president was seen as a remedy for all the ills of the other possible systems. In a temporary body, electors would not be able to exert undo influence on the president. Further, because they met only once, and then in separate states, the threat of foreign influence and local cabals was eliminated. It also was hoped that the electors would be wiser than the masses. Supreme Court Justice Joseph Story later wrote, "A small number of persons, selected by their fellow citizens from the general mass for this special object, would be most likely to possess the information, and discernment, and independence, essential for the proper discharge of this duty."

Once it was decided that electors would select the president, the final hurdle was to figure out how many electors each state should get, so as to protect the southern and smaller states. The first proposal was that states with less than 200,000 inhabitants would get one elector, those with between 200,000 and 300,000 would get two, and those with more than 300,000 would get three. James Madison opposed this, noting that eventually all states would be allotted three votes. A better ratio, he suggested, "varied as that it would adjust itself to the growing population of the States." It was later proposed that each state should have the same number of electors as it had members in the House of Representatives. Finally, the convention agreed on the current plan, in which the number of electors for a state equals its number of representatives in the House of Representatives plus its number of senators.

Because of the three-fifths compromise (discussed in the Preface), the slave states were able to gain valuable electors through this plan, despite their large nonvoting population. Smaller states were benefited even more, since the addition of the two senatorial electoral votes gave them a disproportionately larger share of the electoral college. In 2000, for example, while California had more than 65 times as many people as Wyoming, it only had 18 times as many electoral votes.

The concept that having electors will shield our nation from the folly of its people is ludicrous. The unknown ciphers who serve as electors do not possess, let alone utilize, any special information or discernment lacking in the general population. Placing the electors between the people and the president serves no useful function but merely permits whispers of intrigue and conspiracy as rumors circulate of unfaithful electors voting against their home state's wishes. The elimination of electors from our system would cause no discernable harm.

By contrast, there is much to be said in favor of the allocation of electoral votes. Its sordid legacy as a structural protection for slavery does not resolve the question of whether the system is currently beneficial. The most powerful argument against the electoral college today is that it is antidemocratic, permitting a candidate with fewer popular votes to win the White House. In addition to the 2000 election, this problematic outcome occurred three times in the nineteenth century: when John Quincy Adams was victorious over Andrew Jackson in 1824, when Rutherford B. Hayes beat Samuel Tilden in 1876, and when Benjamin Harrison bested Grover Cleveland in 1888. Critics of the electoral college point out that there have been several other elections in which the "correct" shift of a few thousand votes also could have elected a president who lost the popular vote. A shift of fewer than 10,000 votes in Ohio and Hawaii, for example, would have elected Gerald Ford over Jimmy Carter in 1976, despite Carter's having received 1.6 million more votes.

The danger that the candidate with fewer popular votes will win the electoral college is enhanced by the winner-take-all system. If you win California with 50.1 percent of its vote, you receive 100 percent of its 54 electoral votes. Interestingly, the winner-take-all feature is *not* required by the Constitution but is the choice of the individual states. Only two states, Maine and Nebraska, have chosen to award electors on the basis of vote totals within individual congressional districts.

Whenever voters are divided into distinct regions, there is a risk of an undemocratic result. There is always the chance that the majority of overall voters will differ from the majority in each region, because two razor-thin victories count for more than one overwhelming victory. Thus, a candidate who carries the right states, even by a small margin, can be elected president over an opponent who has carried fewer (or smaller) states by a large margin. Direct election of a president, by contrast, would guarantee that the candidate with the most popular votes prevails.

= = =

The electoral college seems to create undemocratic results in close elections. Yet history and mathematics reveal that the system has significant, if largely unplanned for, benefits. Most notably, the electoral college requires any successful presidential candidate to win in numerous different parts of the country; a dominant base of localized support cannot suffice. Thus, the more popular candidate will lose unless he or she also has geographically widespread support.

Physicist Alan Natapoff compared this result to the need for a baseball team to win four different games in order to capture the World Series. The champion is not the team that scores the most runs during the series but the one that prevails in the most games. In 1996, the Atlanta Braves outscored the New York Yankees 26 to 18, but because their runs were bunched mostly in two games, they lost the series four games to two. In the electoral college, each state can be thought of as a distinct "game." It certainly is possible to outscore your opponent overall (by votes or by runs) and still lose if you do not win enough individual contests. A strong argument can be made that the candidate who has prevailed in more regions of the country will be better able to unify and lead the entire nation.

Natapoff has also argued that dividing voters into districts, as does the electoral college, increases the voting power of each individual. If one considers the power of an individual vote to be "the probability that the outcome will turn on it," it is obvious that the individual vote will prevail only in an extremely tight election race. Such a small margin of victory is incredibly unlikely for an electorate as large as that which votes for the president of the United States. The probability that such a close race will occur in a given state and that such a state will tip the electoral college is much larger. In the 2000 election, for example, the final official margin in Florida was only 537 votes out of the almost 6 million cast.

Natapoff proved that districting (rather than election by popular vote) improves the chance that a single voter will tip an election. What this means, however, is that the increase in individual voting power increases the possibility that the majority candidate will lose.

There is one indisputable bias in the electoral college, which generally has not been noted but which arguably has been one of the most important protectors of domestic tranquility for two centuries. I refer to the bias in favor of politically integrated over politically homogeneous states. Because of the need to win many different states, candidates are forced to concentrate on contested rather than one-sided states. If all that mattered were raw vote totals, the candidates would likely spend more time in those states where they had the greatest probability of locating the most supporters. In the districted election, the candidates must focus more on convincing those in divided regions; in raw vote elections, they can focus more on turning out their followers. For example, in the closing days of the 2000 presidential election, both George W. Bush and Al Gore chose to campaign in battleground states, such as Florida, Wisconsin, Michigan, and Missouri, rather than in the states in which they were leading by comfortable margins.

Campaigning in a politically integrated region is quite different than campaigning solely in front of those who are already in your camp. One must be far more sensitive in trying to persuade those who have sympathies for both sides of a dispute than when one is trying to arouse the faithful. Because a successful candidate under our current system cannot simply go where the votes are, extremism is discouraged. This also means that candidates will tend to ignore the states they have locked up. But it is better for the nation as a whole to have its leaders address communities where people on all sides of the political spectrum reside, rather than only proponents of one extreme or the other. This might produce two candidates who are fighting for the center and thus create a feeling among many voters that "they're all the same." But over the long course of history, it well may have provided a moderating influence, which reduced the likelihood that the White House would be captured by a destructive extremist.

## The Impossible Dilemma of Multicandidate Elections

There is one area in which the electoral college creates a significant risk, not of extremism but of turmoil: a multicandidate contest where no one candidate succeeds in obtaining a majority of the electoral votes. According to the Constitution, if no majority winner emerges, the election is thrown into the House of Representatives, which then decides among the three candidates with the most electoral votes.

The few times the House has been involved in the selection of a president, the results have been abysmal. In 1801, the Federalist-dominated House of Representatives took six days and 36 ballots to select Thomas Jefferson as president over Aaron Burr. In 1824, the House selected John Quincy Adams, even though Andrew Jackson had received more electoral as well as more popular votes. And finally, in the disputed election of 1876, Rutherford B. Hayes made one of the most fateful deals in U.S. history, agreeing to pull federal troops out of the South and end Reconstruction in return for support by the Democrat-controlled House. This is not a history that fills one with confidence.

Unfortunately, whenever there is a strong third-party candidate, there is the possibility that history will repeat itself. In 1968, a shift of just 40,000 votes in Alaska, Missouri, and New Jersey would have denied Richard Nixon his electoral college majority and sent the election to the House of Representatives.

The problem of resolving elections with more than two candidates runs far deeper than the electoral college. In fact, the problem of multicandidate elections is among the most frustrating for anyone hoping to design a democratic system. Because majority rule means at least 50 percent plus 1, only a two-person race guarantees that a majority winner will result. In a contest involving three or more candidates, the votes often will be divided so that no one candidate captures half of the vote. In presidential elections, there have been 18 times, including 1992, 1996, and 2000, when more than half of the voting public voted *against* the winning candidate.

There is no obviously democratic solution for dealing with an election without a majority choice. Simply awarding the victory to the candidate with the most votes, the plurality choice, is dangerously unacceptable. Imagine an election with three candidates: one on the right-to-moderate side of the political spectrum, another on the left-to-moderate side, and a charismatic extremist (think George Wallace in 1968). Assume that the moderate voters total 60 percent and would strongly prefer that the extremist lose. If the moderate vote is divided evenly and plurality rules, then the extremist would win with 40 percent, despite being the choice most opposed by a majority of voters.

≥ ≥ ≥

Even a runoff system is imperfect. The most common problem with a plurality and runoff system is that the most-preferred candidate, the consensus candidate, can lose, despite being preferred by a majority of voters to the eventual winner. Consider this scenario involving three candidates, from the Democrat, Republican, and Green Parties. Sixty voters rank their preference for each candidate, from first choice to third, and a tally of their votes reveals the following distribution (the numbers at the top of each column represent the number of voters who share a particular ranking):

| **Preference Ranking** | **Number of Voters** | | | |
|---|---|---|---|---|
| | *23* | *19* | *16* | *2* |
| *1st Choice* ➤ | Dem | Rep | Green | Green |
| *2nd Choice* ➤ | Green | Green | Rep | Dem |
| *3rd Choice* ➤ | Rep | Dem | Dem | Rep |

Here, the first-place votes are divided so that Dem (the Democratic Party candidate) receives 23, Rep 19, and Green 18. Thus, Green would not be involved in the runoff between the two highest finishers, and Rep would defeat Dem in a runoff 35 to 25. Notice, though, that if you asked the voters whether they preferred Rep or Green, a majority would choose Green by a wide margin of 41 to 19. Similarly, Green also would triumph if the voters were asked to choose between Dem and Green, this time by a margin of 37 to 23. Thus, perversely, the runoff election will exclude the one candidate who would defeat each of the other two in head-to-head competition.

÷ ÷ ÷

The use of head-to-head competition in multicandidate elections was first proposed by the French mathematician Marie-Jean-Antoine-Nicolas de Caritat, Marquis de Condorcet. Condorcet, who was a friend of both Thomas Jefferson and Benjamin Franklin, dreamed of inventing a social mathematics that would create a government and society ruled by reason. Perhaps ironically, perhaps inevitably, he died after being imprisoned during the reign of terror. As has been noted, "Condorcet failed to take numerous other factors into consideration. Not the least of these was the logic of the guillotine."

Condorcet's most lasting contribution was his proposal for dealing with multiple-candidate elections. To say that Condorcet's writing was not initially well received is a considerable understatement:

> *The obscurity and self-contradiction are without any parallel, so far as our experience of mathematical works extend. . . . [N]o amount of examples can convey an adequate impression of the evils. We believe that the work has been very little studied, for we have not observed any recognition of the repulsive peculiarities by which it is so undesirably distinguished.*

Eventually, though, the Condorcet plan became accepted as one of the more plausible attempts to deal with elections among multiple candidates. Today, the principle that the winning candidate of a crowded field should be the one who obtains a simple majority over every other candidate in two-person, head-to-head competition is often called the Condorcet criterion, and the candidate who prevails is called the Condorcet winner.

Yet the Condorcet strategy of head-to-head competition is not without its problems. The most serious weakness is that it frequently will not produce a winner. This situation can be seen in the following simple election. Three people, Alice, Bob, and Charles, belong to a group that is trying to decide whether to endorse the Democratic, Republican, or Green parties. They rank their preferences in the following order:

| Preference Ranking | *Alice* | *Bob* | *Charles* |
|---|---|---|---|
| *1st Choice* ➤ | Dem | Rep | Green |
| *2nd Choice* ➤ | Green | Dem | Rep |
| *3rd Choice* ➤ | Rep | Green | Dem |

Each of the three parties receives one first-place vote, so there is no simple majority winner. Unfortunately, the Condorcet criterion will not help either. In a head-to-head contest, Dem beats Green 2 to 1. In a similar matchup, Green beats Rep 2 to 1. Thus, the group prefers Dem to Green and Green to Rep. One would expect, therefore, Dem to be favored over Rep as well. (If 3 is greater than 2, and 2 is greater than 1, we naturally expect 3 to be greater than 1 also.) The difficulty arises because Rep actually defeats Dem by a 2 to 1 margin.

This sort of confusion would be considered irrational were

such a convoluted set of preferences to be announced by a single individual. Yet this type of irrationality is what has resulted from the group decision process utilizing the Condorcet criterion. It is known as cycling because every winner also loses.

The Reverend Charles L. Dodgson, who gave us the syllogism story in Chapter 1, wrote extensively about voting theory and argued that cycling would become an increasingly likely result as the number of competing candidates increased. To the defenders of the Condorcet criterion, Dodgson scornfully wrote:

> *I am quite prepared to be told . . . "Oh,* that *is an extreme case: it could never really happen!" Now I have observed that this answer is always given instantly, with perfect confidence, and without any examination of the proposed case. It must therefore rest on some general principle: the mental process being probably something like this—"I have formed a theory. This case contradicts my theory.* Therefore *this is an extreme case, and would never occur in practice."*

+ + +

Another French mathematician, Jean Charles de Borda, proposed a different technique for deciding elections with multiple candidates. His plan was to calculate not only a voter's first choice but the voter's ranking of all the candidates. He would have each voter rate all candidates, with 0 points for the least-favored candidate, 1 for the next, and so on through the preferred candidate. Next, the score for each candidate would be determined by adding the points from each voter. The candidate with the highest total score would be deemed the winner.

The Borda ranking eliminates cycling and does reflect the preference of voters considering all of the candidates. Imagine the following preference ranking of three candidates:

| Preference Ranking | Number of Voters | | |
|---|---|---|---|
| | *8* | *7* | *6* |
| *1st Choice* ➤ (2 points) | Dem | Rep | Green |
| *2nd Choice* ➤ (1 point) | Rep | Dem | Rep |
| *3rd Choice* ➤ (0 points) | Green | Green | Dem |

While a plurality chooses Dem as a first choice, Rep is either the first or second choice of every voter. Under the Borda ranking system, Rep wins with 28 points, Dem gets 23 points, and Green, the least favored, gets 12.

One serious problem with the Borda ranking is that it permits manipulation of the outcome via strategic, also known as insincere, voting. In the above example, supporters of the two top finishers, Dem and Rep, might decide to change their preference ranking from one that reflects their true beliefs to one that lowers the ranking of their strongest opponent. The result of such strategic voting would look like this:

| Preference Ranking | Number of Voters | | |
|---|---|---|---|
| | *8* | *7* | *6* |
| *1st Choice* ➤ (2 points) | Dem | Rep | Green |
| *2nd Choice* ➤ (1 point) | Green | Green | Rep |
| *3rd Choice* ➤ (0 points) | Rep | Dem | Dem |

Now Dem has dropped to 16 points, Rep is down to 20, and Green, the least favored, is the new winner with 27 points. As Duncan Black pointed out, the strategy of placing the strongest opponent on the bottom of the list, "would be a great advantage to candidates of mediocre merit, for while getting few top places, they would get few lowest places."

≠ ≠ ≠

The danger of strategic voting was also raised at the Constitutional Convention. In the original electoral college, each elector voted for two candidates; the candidate with the greatest number of electoral votes would be president, and the runner-up would become vice president. (This was changed in 1804 by the twelfth Amendment to the current system under which the president and vice president are selected in distinct votes by the electoral college.) James Madison supported the original plan, because each elector had to select at least one candidate who came from a different state than the elector. Madison predicted that the second choice of each elector would be the best candidate, because the first vote would likely go to a candidate from the elector's home state. Madison discounted the danger of strategic voting:

> *The only objection which occurred was that each Citizen after having given his vote for his favorite fellow Citizen, would throw away his second on some obscure Citizen of another State, in order to ensure the object of his first choice. But it could hardly be supposed that the Citizens of many States would be so sanguine of having their favorite elected, as not to give their second vote with sincerity to the next object of their choice.*

Borda was fully aware of the danger of strategic voting. He also dismissed such concerns, with an almost touching naivete: "My scheme is only intended for honest men."

× × ×

But is there a scheme for all voters, honest and otherwise? In 1951, economist and future Nobel Prize winner, Kenneth Arrow proved that the answer was no, that every possible voting scheme for choosing between more than two candidates can produce an irrational and undemocratic result. Arrow's General Possibility Theorem, which is more commonly called his impossibility theorem, proves that no possible voting system can prevent cycling and simultaneously meet the following four, relatively simple goals:

1. If every person in the group prefers one choice to another, so does the group. (Unanimity)
2. No individual is able to act as a dictator, a person whose vote determines the outcome regardless of the preferences of the other group members. (Nondictatorship)
3. When comparing choices, individuals are free to prefer these choices in any order. (Range or citizens' sovereignty)
4. If an individual or group has a preference between two options, the preference between those two will not be affected by the consideration of other options. In other words, if I prefer Dem to Rep, and Green becomes a new option, I will continue to prefer Dem to Rep. (Independence of irrelevant alternatives)

Arrow's theorem has been summarized as follows: "No voting rule which allows voters to express their true preference and which treats each preference as equally decisive can assure us that it will produce a single preferred choice for three or more voters who have at least three alternatives." The implications of this theorem are staggering. There can be no perfect democratic system for selecting among more than two candidates.

= = =

Every proposed system has its strengths and weaknesses. If we ask which of several candidates has the best claim to victory, we may find several plausible answers. Consider the following situation with four parties, Democratic, Republican, Green, and Reform:

| Preference Ranking | Number of Voters | | | | |
|---|---|---|---|---|---|
| | *18* | *12* | *9* | *4* | *2* |
| *1st Choice* ➤ | Dem | Rep | Reform | Green | Green |
| *2nd Choice* ➤ | Reform | Green | Green | Rep | Reform |
| *3rd Choice* ➤ | Green | Reform | Rep | Reform | Rep |
| *4th Choice* ➤ | Rep | Dem | Dem | Dem | Dem |

If we are using a plurality system, the Democrat wins with the most first-place votes (18). If we choose a runoff system between the two highest finishers, the Republican defeats the Democrat in the runoff 27 to 18. If we use the Borda ranking, the Green Party candidate wins with a total of 78 points. Finally, the Reform Party candidate is the Condorcet winner, defeating all three opponents in head-to-head competitions. Thus, four reasonable, neutral systems can produce four different winners.

≥ ≥ ≥

Is there really one best system? Returning to our discussion of the electoral college, for example, there are undoubtedly plans for selecting a president that resolve some of the problems of the current system. However, those in favor of some alternative proposal should proceed with the following two thoughts in mind. First, the electoral college generally has managed to avoid producing extremist presidents over the course of the nation's history. Second, any new system will of necessity possess its own intrinsic, irremediable flaws.

The ultimate lesson, then, may be that one should be exceedingly modest in making demands on, and claims for, a voting system. As mathematician John Allen Paulos explains,

> *How we should be democratic is the substantive question, and an open experimental approach to this question is entirely consistent with an unwavering commitment to democracy. Politicians who are the beneficiaries of a particular and parochial electoral system naturally wrap themselves in the mantle of democracy and need to be reminded occasionally that this mantle can come in different styles, all of them with patches.*

# 3

# The Positive Value of Consensus

**[T]he right of expulsion [is] too important to be exercised by a bare majority. . . .**

***—James Madison, Constitutional Convention, August 10, 1787***

For any group in which decisions must be made, the ultimate question is how to resolve disagreement. While majority rule is the norm, the framers of the Constitution believed that a slight majority, indicating a closely divided population, is insufficient for certain very important actions. Even in a perfect democracy, in which all members have an equal voice, there are times when 49 percent dissent indicates that perhaps the group is too splintered to make a fundamental change in structure or to pursue a dangerous course. The pejorative the framers used to refer to such a situation was a "bare majority."

Normally, the losers of a vote are expected to accept the outcome willingly. In 1762, Jean-Jacques Rousseau wrote that a promise of consent to majority rule is incorporated into the initial consent to the social contract: "Except this original contract, the vote of a majority always binds the rest, this being a result of the contract itself."

But Rousseau overstates the power of the majority. What Benjamin Cardozo wrote of contracts in general, applies to the social contract as well: "We are not to suppose that one party was to be placed at the mercy of the other." The framers were well aware that simple majority rule would empower a majority "to sacrifice to its ruling passion or interest both the public good and the rights of other citizens."

One crucial safeguard was the requirement that for issues of great importance or great delicacy, a simple majority should not suffice. As the Supreme Court declared in 1943, the entire philosophy of constitutionally protected liberties is premised on the belief that some interests are to be protected *from* majority decision making:

> *The very purpose of a Bill of Rights was to withdraw certain subjects from the vicissitudes of political controversy, to place them beyond the reach of majorities and officials and to establish them as legal principles to be applied by the courts. One's right to life, liberty, and property, to free speech, a free press, freedom of worship and assembly, and other fundamental rights may not be submitted to vote; they depend on the outcome of no elections.*

Equally important, the framers believed that there are certain times when action is unwise unless backed by a consensus. *Consensus* in this context means an indication of strong community agreement, where a decision can be seen to reflect a sense of harmony and unity. For the framers, the mathematical translation of *consensus* is the requirement that a decision be made by at least two-thirds of those voting.

In the original Constitution there were five different provisions requiring a two-thirds vote for action to be taken. Two-thirds of both houses are necessary to override a presidential veto and to propose amendments to the Constitution. Two-thirds of either house is necessary for expelling a member. Finally, two-thirds of the Senate is required both for conviction after impeachment by the House of Representatives and for approval of treaties.

The choice of two-thirds to represent consensus was not made haphazardly. The fraction was chosen only after extensive deliberation and consideration of alternatives, ranging from requirements of a simple majority, to a three-quarters supermajority, to unanimity. Except for requiring that three-fourths of the states approve any constitutional amendment, the number that was used by the framers for decisions requiring a national consensus was two-thirds. No number can represent in literal form a concept as intangible as consensus. Nonetheless, two-thirds was settled on because it was deemed to be sufficiently more difficult to achieve than a simple majority, yet not so onerous as to be practically unobtainable.

÷ ÷ ÷

The framers considered the power of Congress to remove a popularly elected president as essential to prevent corruption and the abuse of executive power. Yet they understood that this important power could be used safely only if there were a consensus regarding its necessity in any particular instance. They did not want removal to be either too easy or too difficult. An early proposal to permit removal of the president by a simple majority vote of both houses was replaced by the current provision, which authorizes impeachment by a majority of the House of Representatives but requires a two-thirds vote in the Senate for conviction.

At the ratifying conventions, the two-thirds requirement was cited as evidence of the care with which power was distributed by the Constitution: "[N]o offender can escape the danger of punishment. Officers, however, cannot be oppressed by an unjust decision of a bare majority; for . . . no person shall be convicted without the concurrence of two thirds of the members present." Alexander Hamilton similarly argued that requiring a two-thirds vote for removal would provide a "security to innocence."

Later, Justice Story would extol the virtues of this two-thirds requirement:

*If a mere majority were sufficient to convict, there would be danger, in times of high popular commotion or party spirit, that the influence of the House of Representatives would be found irresistible. The only practicable check seemed to be, the introduction of the clause of two thirds, which would thus require a union of opinion and interest, rare, except in cases where guilt was manifest, and innocence scarcely presumable.*

Although extremely unpopular, President Andrew Johnson avoided conviction by two-thirds of the Senate by a single vote, with 35 senators voting to convict, and 19 voting to permit him to remain in office. While the merits of President Johnson's tenure are still being debated, it is surely noteworthy that the primary ground for the impeachment proceeding was his firing of Secretary of War Edwin Stanton in violation of the Tenure of Office Act of 1867, a law which was subsequently found to be an unconstitutional restriction on the power of the president.

President William Jefferson Clinton was impeached easily by the House of Representatives. In a virtually straight party-line vote, with House Republicans voting 223 to 5 for at least one article of impeachment and Democrats voting 201 to 5 against, a simple majority was obtained. But in the Senate, made up of 55 Republicans and 45 Democrats, Republicans did not control two-thirds of the vote. In fact, no political party has controlled two-thirds of the Senate since the Civil War. Thus, the constitutional numerical requirement has meant that the removal of a president by the legislative branch is impossible unless members of that president's own party are convinced his offenses are so grave as to require conviction. The Constitution as a practical matter prevents party-line removal of a president.

The framers knew that obtaining a two-thirds consensus vote often would be difficult. Those who survive impeachment votes can find comfort in the words of Justice Story: "[I]f the guilt of a public officer cannot be established to the satisfaction of two thirds of a body of high talents and acquirements, which sympathizes with the people and represents the states[,] . . . it must be that the evidence is too infirm, and too loose to justify a conviction."

Just as impeachment gave the legislative branch power over the executive, the framers believed that the executive required some sort of veto power over acts of Congress. As future Supreme Court Justice James Iredell declared at the North Carolina Ratifying Convention, some protection was needed for those times, "where a bare majority has carried a pernicious bill. . . ." According to James Madison, the two purposes of the veto were to "defend the Executive Rights" and "to prevent popular or factious injustice."

The first proposal for presidential veto power was that the chief executive should have "an absolute negative," for "[w]ithout such a self-defence the Legislature can at any moment sink it into non-existence." The idea that the president should be given the power to defeat single-handedly actions of the legislature was quickly rejected. Roger Sherman argued that "No one man could be found so far above all the rest in wisdom," and George Mason warned that an absolute negative "paved the way to hereditary Monarchy." Finally, Madison suggested that requiring a greater percentage of each house to override a veto than to pass legislation would enable the legislature to prevent abuses of the presidential check on legislative power: "[I]f a proper proportion of each branch should be required to overrule the objections of the Executive, it would answer the same purpose as an absolute Negative."

The debate then focused on what that "proper proportion" should be: two-thirds or three-quarters. At one point in the proceedings, the delegates agreed on requiring a three-fourths vote to override, because, in the words of Gouverneur Morris, "The excess rather than the deficiency of laws was to be dreaded."

Others objected that the three-quarters provision "puts too much in the power of the President." The fear was that if only one-quarter of either house could uphold a veto, "a few Senators having hopes from the nomination of the President to offices, will combine with him and impede proper laws." To the argument that there was no real difference between two-thirds and three-quarters, George Mason snidely remarked, "little arithmetic was necessary to understand that 3/4 was more than 2/3."

James Madison neatly summed up the issue: "We must," he said, "compare the danger from the weakness of 2/3 with the danger from the strength of 3/4." Although Madison, with George Washington's assent, argued that "on the whole, the former was the greater," the Convention voted to reduce the supermajority necessary for an override to two-thirds. As Justice Story was later to remark, "[A]n expression of opinion by two thirds of both houses in favor of a measure certainly affords all the just securities, which any wise, or prudent people ought to demand in the ordinary course of legislation. . . ."

The Constitution, therefore, demands that there be a reasonably strong consensus in the Congress before a presidential veto can be overridden. For example, in 1992, a popular bill to regulate cable television rates became law over President George Bush's veto, by vote of 74 to 25 in the Senate and 308 to 114 in the House of Representatives. By contrast, President Clinton's 1996 veto of a bill to prohibit certain late-term abortions, so-called partial birth abortions, was sustained, when the Senate could muster only 57 votes in favor of an override. Thus, even though a majority of the Senate voted for the law, and more than two-thirds of the House voted to override the law (by a vote or 285 to 137), there was an insufficient congressional consensus to reverse the veto of the president.

≠ ≠ ≠

The framers of the Constitution also believed that the power to make treaties was too important to leave to a bare majority. Entanglements with foreign governments were considered to be dangerous and unnecessary. As George Washington declared in his Farewell Address, "'Tis our true policy to steer clear of permanent Alliances, with any portion of the foreign world."

In that same spirit, the Constitution requires a two-thirds approval of the Senate to ratify a treaty signed by the president. (The reason for giving this power to the Senate, whose members then were selected by state legislatures, was to protect the interests of the individual states in the making of foreign policy.) As with the veto-override provision, the decision to require two-thirds was not made without extensive debate. John Jay wrote that requiring a two-thirds mandate for treaties would provide powerful insurance against foreign corruption:

> *He must either have been very unfortunate in his intercourse with the world, or possess a heart very susceptible of such impressions,*

*who can think it probable that the President and two thirds of the Senate will ever be capable of such unworthy conduct. The idea is too gross and too invidious to be entertained.*

At the Constitutional Convention, some had argued that the two-thirds requirement was unnecessarily onerous. James Wilson complained that "If the majority cannot be trusted, it was a proof . . . that we were not fit for one Society." He argued that a two-thirds requirement permits "the power of a minority to controul the will of a majority." Nonetheless, his proposal that treaties be ratified by only a majority of the Senate was defeated. As delegate Hugh Williamson explained, "This will be less security than 2/3 as now required."

In *The Federalist Papers,* Alexander Hamilton argued forcefully that security was to be found in numerical protection:

*The security essentially intended by the Constitution against corruption and treachery in the formation of treaties, is to be sought for* in the numbers *and characters of those who are to make them. The JOINT AGENCY of the Chief Magistrate of the Union, and of two thirds of the members of a body selected by the collective wisdom of the legislatures of the several States, is designed to be the pledge for the fidelity of the national councils in this particular.*

Thus, the Constitution provides security against ill-considered or ill-motivated treaties by requiring approval by consensus. A bare majority is not sufficient to bind the United States with other nations.

Since the ratification of the Constitution, ten treaties have received a majority vote in the Senate but have failed to receive the requisite two-thirds vote. The most prominent of these was the Treaty of Versailles, which would have brought the United States into the League of Nations. After rejection of the Treaty of Versailles, the Senate became known as the "grave-yard of treaties." The last time a major treaty was rejected despite majority senatorial support occurred in 1935, when the Senate, with a vote of 52 yeas to 36 nays (8 senators abstaining), rejected a treaty committing the United States to abide by decisions of the World Court.

The number of treaties so rejected understates the significance of the two-thirds requirement. As Senate historian George Haynes wrote, "[T]he rule's most calamitous effects are psychological." There is no way to know how many treaties either were altered in negotiations or were never submitted under the pressure of obtaining approval by the requisite margin.

Not surprisingly, the supermajority requirement has come under repeated criticism, such as that offered in George Haynes's observation:

> *[I]t heartens any tiny group having a direct interest adverse to a pending treaty to attempt by delays and bargaining to persuade enough colleagues to join them to make up a "recalcitrant one-third plus one." Such an* ad hoc *bloc . . . can and does exercise a "pathological obstruction" in the handling of our foreign relations such as is exercised by so small a minority in no other legislative body in the world.*

The constitutional "obstruction," however, has been largely eliminated by the use of executive agreements. While there are many permutations, the two major forms of executive agreements are sole executive agreements, those which the president enters into without congressional approval, and congressional-executive agreements, those with foreign countries which become binding on majority vote of the Senate and the House of Representatives. Although not specified in the Constitution, these agreements generally have replaced treaties as the United State's predominant form of international agreement. Between 1960 and 1983, for example, there were 344 treaties and 6,448 executive agreements, meaning that just 5 percent of these U.S. foreign commitments were formalized by treaty.

The validity of sole executive agreements, while controversial, has a solid constitutional underpinning. The Constitution recognizes "the very delicate, plenary and exclusive power of the President as the sole organ of the federal government in the field of international relations." Additionally, it can be argued that certain international crises can be resolved only by immediate unilateral presidential action.

The constitutional support for congressional-executive agreements is far less certain. As Louis Henkin wrote in his classic work, *Foreign Affairs and the U.S. Constitution,* "Constitutional doctrine to justify Congressional-Executive agreements is not clear or agreed." Nonetheless, as Professor Henkin also notes, "Neither Congress, nor Presidents, nor courts, have been seriously troubled by these conceptual difficulties and differences. Whatever their theoretical merits, it is now widely accepted that the Congressional-Executive agreement is available for wide use, even general use, and is a complete alternative to a treaty."

The reason for its appeal is obvious. Because this agreement need only be approved by a simple majority of both houses, presidents need not fear being blocked by a vote of one-third plus one of the senators present. As one historian has noted, "On certain occasions, when the treaty-making method has failed or seemed likely to fail, [the President] has accomplished his purpose by substituting the more facile type of agreement."

Some have said that not only are congressional-executive

agreements interchangeable with treaties but questions as to their constitutionality have become "unaskable." Indeed, despite the frequency of such agreements, their constitutionality has never been ruled on directly by the Supreme Court. A powerful argument can be made that the validity of congressional-executive agreements will be considered a political question, not subject to judicial review.

Nevertheless, questioning whether congressional-executive agreements should be viewed as interchangeable with treaties ought not be dismissed as "constitutional purism." If such agreements are permitted to replace treaties, we will lose something important. As John W. Foster, a former U.S. secretary of state, noted in 1906,

> *[I]n requiring that all treaties should secure the vote of two-thirds of the Senate, the framers of the Constitution emphasized their conviction that the Executive should enter into no stipulations with a foreign power, which did not command the support of a large majority of the people of the United States.*

If congressional-executive agreements replace treaties, we lose the guarantee that a large majority, rather than just a bare majority, supports the links made between the United States and the rest of the world. We will forfeit what Alexander Hamilton referred to as "the advantage of numbers in the formation of treaties."

# 4

# The First Veto

**No invasions of the constitution are so fundamentally dangerous as the tricks played on their own numbers. . . .**

***—Thomas Jefferson***
***"Opinion on the Bill Apportioning Representation" (1792)***

On April 30, 1789, George Washington was inaugurated as the first president of the United States. On April 5, 1792, at the behest of Thomas Jefferson, President Washington issued the first presidential veto of a bill passed by Congress. At the heart of the dispute—fractions.

The question was how to calculate the number of members each state should send to the House of Representatives. The Constitution requires that representatives "shall be apportioned among the several States according to their respective Numbers. . . ." The problem, though, is what to do when the numbers do not come out evenly.

First, let's look at the ideal situation for finding the number of representatives for each state:

1. Determine the desired total number of House members.
2. Divide that number by the nation's population to determine the overall ratio of persons per representative.
3. Multiply that ratio by each state's population.

Mathematicians refer to this quantity as each state's quota or "fair share."

Imagine that the United States began as a tiny nation of 2,000 people, with 3 states: Virginia had a population of 900, Vermont 500, and Massachusetts 600. If those designing the national legislature determined that an appropriate size would be 60 members, each state's share could be determined by the formula

$$\frac{\text{Total number of seats}}{\text{National population}} \times \text{state population} = \text{number of seats for state}$$

Virginia $\frac{60}{2{,}000} \times 900 = 27$

Vermont $\frac{60}{2{,}000} \times 500 = 15$

Massachusetts $\frac{60}{2{,}000} \times 600 = 18$

Thus, Virginia would have 27 representatives, Vermont 15, and Massachusetts 18. This worked out neatly because all the numbers were selected so that the answer for each state was a whole number. But in the real world, especially as populations grow, the chances of being so lucky are minuscule. Just imagine that two people leave Virginia, one for Vermont and one for Massachusetts. Suddenly, the equations do not work out so well:

Virginia $\frac{60}{2,000} \times 898 = 26.94$

Vermont $\frac{60}{2,000} \times 501 = 15.03$

Massachusetts $\frac{60}{2,000} \times 601 = 18.03$

The ideal number for each state is no longer a whole number. No apportionment scheme could give each state its precise quota. As mathematician Paul Hoffman notes, "The problem is that although the loyalties of a congressman can be divided, his body cannot be; human beings, like pennies or electric charges or subatomic states, are quantized."

This was the problem facing Congress in 1791. According to the first census, the state-by-state populations were

| | |
|---|---|
| Connecticut | 236,841 |
| Delaware | 55,540 |
| Georgia | 70,835 |
| Kentucky | 68,705 |
| Maryland | 278,514 |
| Massachusetts | 475,327 |
| New Hampshire | 141,822 |
| New Jersey | 179,570 |
| New York | 331,589 |
| North Carolina | 353,523 |
| Pennsylvania | 432,879 |
| Rhode Island | 68,446 |

| | |
|---|---|
| South Carolina | 206,236 |
| Vermont | 85,533 |
| Virginia | 630,560 |
| Total | 3,615,920 |

Obviously, no simple formula would result in a whole number for each state. Moreover, to ensure that even the smallest state is represented, the Constitution requires that each state have at least one member in the House of Representatives. This further complicates the math. After much wrangling, Congress approved a bill, championed by Alexander Hamilton, with the following apportionment for a 120-seat House:

| | |
|---|---|
| Connecticut | 8 |
| Delaware | 2 |
| Georgia | 2 |
| Kentucky | 2 |
| Maryland | 9 |
| Massachusetts | 16 |
| New Hampshire | 5 |
| New Jersey | 6 |
| New York | 11 |
| North Carolina | 12 |
| Pennsylvania | 14 |
| Rhode Island | 2 |
| South Carolina | 7 |
| Vermont | 3 |
| Virginia | 21 |
| Total | 120 |

Thomas Jefferson, then Washington's secretary of state, strongly opposed the bill. The primary mathematical sin of the bill was that it merely gave the number of representatives for each state without disclosing how the numbers were derived. The bill, Jefferson said, did not explain "any principle at all, which may shew it's [*sic*] conformity with the constitution, or guide future appointments." As Jefferson noted, this would permit future Congresses to manipulate apportionment after each census to achieve politically desired outcomes, "according to any . . . crotchet which ingenuity may invent, and the combinations of the day give strength to carry."

Hamilton eventually did explain his plan to President Washington. After finding the ideal number as before, Hamilton's plan ignored remainders and allotted to each state the number of representatives equal to the whole number of its quota. If the total so allocated was less than the desired size of the full House, the remaining seats were allocated one by one to the states with the largest fractional remainders.

Thomas Jefferson proposed a different plan which was numerically more complicated. For any given census, a special number had to be discovered. This number was calculated so that dividing it into each state population separately, ignoring the fractional remainders, and adding the whole numbers of each quotient would total whatever was decided to be the desired total number of representatives. The whole-number quotient obtained for each state would be its allotted number of representatives.

A direct comparison between the results of the Hamilton and Jefferson allotment plans when applied to the first census reveals what this conflict really was about.

| | Hamilton's Plan | Jefferson's Plan |
|---|---|---|
| Connecticut | 8 | 8 |
| Delaware | 2 | 1 |
| Georgia | 2 | 2 |
| Kentucky | 2 | 2 |
| Maryland | 9 | 9 |
| Massachusetts | 16 | 16 |
| New Hampshire | 5 | 4 |
| New Jersey | 6 | 6 |
| New York | 11 | 11 |
| North Carolina | 12 | 12 |
| Pennsylvania | 14 | 15 |
| Rhode Island | 2 | 2 |
| South Carolina | 7 | 7 |
| Vermont | 3 | 3 |
| Virginia | 21 | 22 |

Note that of the 15 states, 11 have the same number of seats under both systems. Under Hamilton's plan, however, the two remaining seats go to smaller states, Delaware and New Hampshire. Under Jefferson's plan, the remaining seats go to larger states, Virginia and Pennsylvania. Such a result is not a quirk of the 1790 census. It turns out that the Jefferson plan is an extraordinarily effective machine for allocating extra seats to large states. An examination of census data between 1790 and 1970 revealed that in every single allocation, that is 19 out of 19 times, the Jefferson plan would favor large states over small states.

Such a result hardly can be considered coincidental. As has

been observed, "being a shrewd man of science, [Jefferson] was no doubt fully aware of it. But it was a favoritism he approved of. . . ." After all, Jefferson was from Virginia, the largest state in the Union. So was the person he was trying to convince, President Washington. Thus, President Washington issued the first veto in U.S. history, and Jefferson's plan was eventually enacted into law, at least for a while.

Since then, the nation has used several other plans, each with its own peculiarities. The Supreme Court finally addressed the issue of whether there are any limitations imposed by the Constitution, on March 31, 1992, less than a week shy of the 200-year anniversary of President Washington's veto. In *U.S. Department of Commerce v. Montana,* the Court upheld a plan known as the method of equal proportions, which had been used since 1941 (see box next page). This plan is far more complicated than both the Jefferson and Hamilton plans, but it has no strong bias between large and small states.

The problem was that it resulted in Montana's having only one congressional district, whose population of 803,655 exceeded its mathematically ideal number by more than 230,000. A different method, known as the method of the harmonic mean, would have split Montana into two districts, each with a population of only 170,000 less than ideal. Of course, Montana's gain must be at someone's expense, and under the alternative method, the State of Washington would have been reduced from 9 to 8 representatives.

Each of the two systems had a mathematical advantage. With the method of the harmonic mean, the total numerical difference between the size of each district and the "ideal" district size was minimized. Under the method of equal proportions, though, the percentage that each district deviated from the "ideal" size was smaller. The Court ruled that "neither mathematical analysis nor

> **The Method of Equal Proportions**
>
> The method of equal proportions works by devising a priority list for allocating all of the seats for the House of Representatives. First, one representative is assigned to each state, to satisfy that constitutional guarantee. Next, the population of each state is divided by a series of divisors, following the pattern $\sqrt{n(n-1)}$. Thus, each state's population is divided by $\sqrt{2}$ (which is $\sqrt{2 \times 1}$), then by $\sqrt{6}$ ($\sqrt{3 \times 2}$), then $\sqrt{12}$ ($\sqrt{4 \times 3}$), and so on. Finally, all the quotients are arranged in a single list. The 51st seat is assigned to the state with the largest quotient, and so on down the list, until the 435th seat is allocated.

constitutional interpretation" indicated that one type of difference was more significant than the other. Thus, the Court said, under the Constitution, it was left to Congress's "good faith choice of a method of apportionment."

The most important lesson from this controversy is an understanding of what "good faith" requires. The system being challenged had been used for half a century, and there was no evidence that it was currently being utilized to "maintain partisan political advantage." Thus, its continued use could be characterized as a good faith decision.

By contrast, the system's initial selection was not so benign. In 1941, the dominant political party was the Franklin Roosevelt–led Democratic Party. The decision to select a new method of allocating House seats clearly benefited the party in power. The immediate effect of choosing the method of equal proportions was that Arkansas (a strong Democratic state) gained one seat at the expense of Michigan (a strong Republican state). One angry Michigan Republican said sarcastically at the time: "One Democrat is quoted—perhaps erroneously—as having stated that there was no partisanship in this bill; that it was merely a measure to give the Democratic Party a Congressman and a Presidential elector which they could not otherwise be certain of securing."

If Michigan had sued in 1941, rather than Montana 50 years later, the result might well have been very different. "Good faith," if it means anything mathematically, must mean not fixing the results before the vote is taken. Mathematical formulas are seductive, because they present an illusion of impartiality and infallibility. But if the result, or likely result, is known when a formula is chosen, it is merely self-serving protectionism masquerading as objective mathematics. Fairness is not guaranteed by the complexity of the equation.

# 5

# What Does Equality Equal?

**A classification having some reasonable basis does not offend against [the Equal Protection Clause] merely because it is not made with mathematical nicety. . . .**

***—Justice Willis Van Devanter***
**Lindsley v. Natural Carbonic Gas Co. *(1911)***

Every political decision has supporters and opponents. Every election has winners and losers. If there is to be a loser in the democratic process, it is appropriate that it be the side with fewer adherents. But, as James Madison warned, democratic rule permits "majority factions" to abuse this power and create tyranny and oppression.

While the Constitution has long been seen as the protector of minorities from such abuse, the modern expression of this philosophy can be traced to a footnote in an otherwise unexceptional 1938 case, *U.S. v. Carolene Products Co.* In *Carolene Products,* the Supreme Court upheld a federal ban on the sale of "filled milk," a milk product with coconut oil substituted for the natural butterfat. In finding that there was no constitutional right to sell filled milk, the Court stated that it would give great deference to "regulatory legislation affecting ordinary commercial transactions." In its now-famous footnote 4, the Court explained why certain other regulations required closer judicial scrutiny: "[P]rejudice against discrete and insular minorities may be a special condition, which tends seriously to curtail the operation of those political processes ordinarily to be relied upon to protect minorities, and which may call for a correspondingly more searching judicial inquiry."

This focus on "discrete and insular minorities" has created much controversy. Professor Bruce Ackerman has pointed out that in the political arena, "discreteness" and "insularity" are generally advantageous, increasing group loyalty, minimizing organizational costs, and permitting an intense single-mindedness which may lead to success over disorganized, diffuse interests. A look at any well-heeled special interest and its army of lobbyists in Washington, D.C., will confirm the accuracy of this insight.

Moreover, if the nation is seen as consisting of many different groups, then minorities can enjoy some of the fruits of majoritarianism by linking with each other or with majority groups to protect their own interests. One need not possess the majority of votes if one can combine with enough voters or simply gain the sympathy or respect of the majority.

The *Carolene Products* concern, therefore, is best understood by focusing on the full phrase, "*prejudice* against *discrete* and *insular minorities*." The four italicized words can be restated as follows: *Prejudice* refers to the situation where individual members of a group are disliked or distrusted not because of their particular personal characteristics but because of their membership in that group. *Discrete* refers to a group that can be identified readily by government, in statute, regulation, or application of laws. A group is *insular* if it can be attacked by the government without harm to others. *Minority* status means that the disfavored group members are not numerous enough to protect themselves at the ballot box.

Judicial scrutiny of democratically enacted laws, then, is needed especially for protecting those unpopular groups which lack political power and can be both identified readily and attacked without causing harm to the larger or more popular segments of society. This certainly is not a complete description of the role of the Court in protecting individual or minority rights but a useful reminder: To fulfill the promise of equality within a democratic society, the existence of prejudice necessitates recourse to a countermajoritarian tribunal.

× × ×

Equality is a fundamental mathematical concept, and yet, ironically, mathematics has only a limited ability to identify what is equal in society. In many cases, numbers fall short either of proving discrimination or of justifying it.

In *Craig v. Boren*, the Supreme Court ruled that an Oklahoma law, which prohibited the sale of certain alcoholic beverages to males under the age of 21 and females under the age of 18, violated the Equal Protection Clause. The state tried to defend its law on the grounds that statistics proved that alcohol consumption by young men posed a greater danger than consumption by young women. Specifically, only 0.18 percent of females in that age group had been arrested for driving under the influence of alcohol, as opposed to 2 percent of same-aged males. The Court held that this disparity was insufficient to justify using gender to distinguish between legal and illegal conduct: "Certainly if maleness is to serve as a proxy for drinking and driving, a correlation of 2% must be considered an unduly tenuous 'fit.'" Justice Stevens wrote in concurrence, "[I]t does not seem to me that an insult to all the young men of the State can be justified by visiting the sins of the 2% on the 98%."

Note, however, that when the drinking age is raised to 21 years for all people, age is serving as a proxy for drinking and driving on the basis of an even more tenuous fit. After all, just barely 1 percent of those in the 18 to 21 age group would have been arrested for driving under the influence, yet the use of age as a "proxy for drinking and driving" did not raise any constitutional concern. Wasn't an insult to all the young people of the state being justified by visiting the sins of the 1 percent on the 99 percent?

The reason that a 1 percent fit justifies an age distinction but a 2 percent fit does not justify a gender discrimination lies in the nonmathematical realm. Gender discrimination creates a special harm, an unacceptable risk that the government is making "judgments about people that are likely to stigmatize as well as to perpetuate historical patterns of discrimination." As the Court recently has held, sex classifications "may not be used, as they once were, to create or perpetuate the legal, social, and economic inferiority of women."

Sex discrimination, therefore, cannot be justified on the grounds that more women or more men are one way or another. As Justice Sandra Day O'Connor has explained "[T]o say that gender makes no difference as a matter of law is not to say that gender makes no difference as a matter of fact." Statistics cannot justify differential treatment based on sex. The Constitution prohibits even statistically sound sex discrimination.

= = =

Using statistics to prove unconstitutional discrimination is a different, but no less troubling problem. A few cases have been remarkably easy. In *Yick Wo v. Hopkins,* for example, a local ordinance barred laundries from operating in wooden buildings without a government permit. All but one white applicant received the permit, and none of the over 200 applicants of Chinese ancestry were successful. The Court found that the numbers proved that the law was being administered "with a mind so unequal and oppressive as to amount to a practical denial by the State of [the] equal protection of the laws. . . ."

Similarly, in *Gomillion v. Lightfoot,* the Supreme Court found a violation of the Fifteenth Amendment right to vote free of racial discrimination, when the state legislature altered a city's boundaries "from a square to an uncouth twenty-eight-sided figure." The effect of this change was to exclude 395 of the town's 400 African American voters, without affecting a single white voter. The Court decried this "essay in geometry and geography" and stated that "the conclusion would be irresistible, tantamount for all practical purposes to a mathematical demonstration," that the legislation was concerned solely with segregating voters by race.

These are exceptional cases, however. In general, when the numbers are not overwhelming, the Court has held that a law which does not identify its beneficiaries specifically by race or gender will not be deemed unconstitutional merely because it has a harsher effect on one race [or gender] than the other: "Disproportionate impact is not irrelevant, but it is not the sole touchstone of an invidious . . . discrimination. . . ." More than numbers are necessary to prove such invidious discrimination.

In *McClesky v. Kemp,* the Supreme Court rejected a broad constitutional challenge to the death penalty that was based on a statistical study of 2,000 murder cases in Georgia. According to the Court, the study showed that the death penalty "was assessed in 22% of cases involving black defendants and white victims, 8% of cases involving white defendants and white victims, 1% of cases involving black defendants and black victims, and 3% of cases involving white defendants and black victims." In short, the study revealed that the death penalty was imposed more frequently when whites were victims and most frequently when

African Americans were convicted of murdering whites. But despite the numerical disparity, all of the justices agreed with the fundamental principle that statistics, by themselves, could not prove that race had been a factor in any particular sentencing decision: correlation does not prove causation.

The issue that divided the Court was whether racial prejudice was the likely reason for the statistical discrepancy. The differing approaches reveal, in a nutshell, much about the debate over race in the United States today. For four justices, the impact of the numbers was very strong; they saw the racially discriminatory outcome as "consonant with our understanding of history and human experience." By contrast, the numerical discrepancy was not viewed as constitutionally significant by those with a more benign view of current race relations. As Justice Lewis Powell stated for a majority of the Court, "[W]e decline to assume that what is unexplained is invidious."

The ultimate decision, then, was not about statistics at all. Instead, the debate turned on the extent to which each justice saw racial discrimination as prevalent in the United States today, and thus a likely explanation of the statistical findings. The question was brought to the fore by statistics, but the meaning of the numbers had to be ascertained in a far less rigorous, far more subjective manner.

## An Imperfect Model for Affirmative Action

There is no arithmetic or mathematical resolution to the conflict over affirmative action. The ultimate, bottom-line value judgment questions are quintessentially nonmathematical. Nonetheless, there is a role for mathematical analysis. Poor reasoning often has clouded and inflamed the debate. A carefully drawn mathematical portrait can help by identifying the areas where people of good faith should agree and can disagree, improving the quality of debate, and even suggesting possible future steps that may unite the two sides.

It sometimes has been all too easy, yet completely unpersuasive, for each side in this debate to declare either that all forms of distinction are equally odious (distinction = discrimination) or that members of different racial groups are never treated equally (treatment of African American ≠ treatment of white). Neither relationship is so simple.

Imagine a foot race on a straight track between two runners. If one runner were given a head start, we would say that runner had received an unfair advantage. But now consider a circular track. In this case, we want the outside runner to start ahead of the other, in order that they run the race over a comparable distance. How to fulfill the imperative of a fair race, then, depends on the shape of the track.

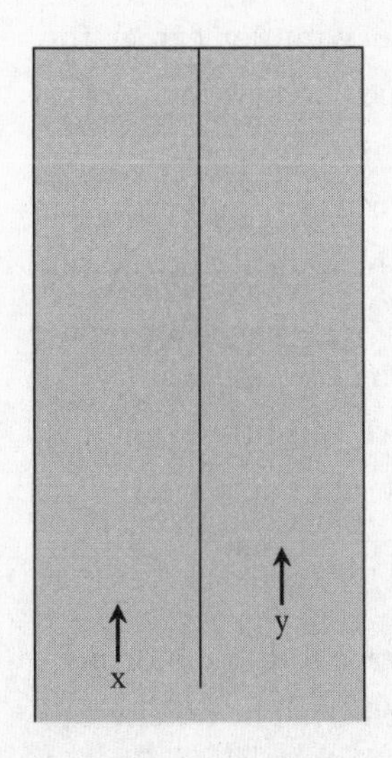


Straight Track

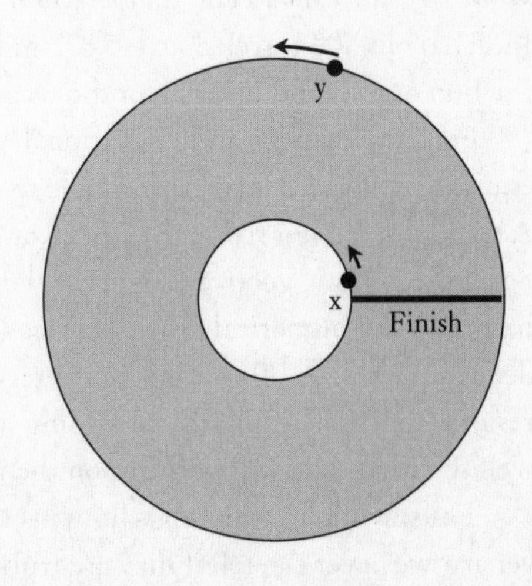


Circular Track

For the affirmative action debate, this distinction equates to the question of whether African American candidates (for jobs, graduate school slots, etc.) face a "longer track" than whites. There should be no disagreement that years after the 1954 ruling of *Brown v. Board of Education* and the Civil Rights Act of 1964, as school systems and employers continued to be found guilty of deliberate racial discrimination, the paths for success were of greatly disproportionate length. One also need not be a Pollyanna to assert that the situation has since improved. The disputed question, of course, is: How much improvement has there been? Are we close enough to the straight track today?

One reason this question is so difficult to answer is that there is no common unit agreed on for measuring the amount of equality in a society. With our track analogy, we have no such difficulty. Distance is readily measured in miles or kilometers, and speed is a basic mathematical concept, relating distance covered to time traveled. We can agree that a given race is fair if the person who travels with greater average speed over the course of the race wins. We all can accept that, in 1997, the winning car at the Indianapolis 500 traveled at 145.827 miles per hour, which was a fraction of a second faster than the second-place finisher.

For employment and educational opportunity, however, the equivalent of speed is not susceptible to mathematical calculation. A standardized test score, the LSAT or MCAT for example, does not reflect how good a lawyer or doctor the applicant will become. The numerical scores serve as mere substitutes for that determination. A 30-minute job interview after a review of a resume functions similarly, providing a mere estimation of how well the candidate will perform on the job.

Thus, while we can tell who wins the race to the job or university, we never can tell if they are truly the best. The quickest to the finish line is not necessarily the fastest if the routes traveled are different. Therefore, a fundamental point in the affirmative action debate is how to determine the analogue of speed, in other words, how to determine who is most qualified.

≥ ≥ ≥

As with any mathematical process, we must begin by defining our terms. Participants in the debate over this issue should be considered of "good faith" if they fit the following three descriptions: honest, nonracist, and nonjingoistic. *Honest* means considering new arguments from each side to have an equal likelihood of being plausible and being willing to be swayed by a persuasive argument from either side. *Nonracist* means believing that no one race is better than any other. *Nonjingoistic* means not believing that one's own group (or race) should triumph merely because of shared group (or racial) characteristics.

In an ideal world, employment and admission decisions would be based purely on merit. Although there would be no quantitative measure for determining which applicant would be most qualified, the decisions would be made under the assumption that some applicants to law school would be better lawyers than those rejected or that some candidates for a given job would be better employees than those turned down. For this analysis, we will term those who would be better to be more qualified.

If you say that it is meaningless to discuss a concept as vague as a better lawyer, it is not because the concept does not exist. If there were no such thing as a better lawyer, then those who apply should be admitted to law school on a first-come, first-served basis. The real problem is that the concept of better is both subjective and indeterminate. Nonetheless, admissions and personnel offices still must make the ultimate decision. Thus, although people of good faith may differ over what makes a particular candidate better, we should agree that whoever is better should prevail.

When a mathematician allows a few variables to represent a complex situation, this representation is called a model. When a lawyer sees one simple set of characteristics (such as test scores) used to stand in for more important but less ascertainable characteristics, this substitution is called a proxy. Both a good model and a good proxy make the complicated simple and permit easy calculation and comparison.

All models, however, are imperfect, as they omit complicating factors. The problem was well described by mathematician John Allen Paulos: "[T]he certainty of mathematical conclusions derived from the model does not always extend to the assumptions, simplifications, and data that one uses to construct it. The latter are sticky, nebulous and quite fallible. . . . [Reality] is indefinitely complex and impossible to capture completely in any mathematical model."

Therefore, we should be able to agree that, notwithstanding the best efforts of all concerned, the fit between the factors being weighed and the ideal of finding the most qualified always will be imperfect. Furthermore, there undoubtedly will be good faith disagreements over the degree of imperfection. If we agree that we want 100 percent of our selected candidates to be qualified, meaning that the last accepted applicant is better than the best rejected candidate, much of the disagreement over affirmative action can be seen as battling over the creation of a model, or the selection of a proxy, used in the hiring and appointments process.

÷ ÷ ÷

To create a mathematical portrait of this battle, we can consider a contest with several players. To do so, we will borrow a form of analysis from the eighteenth-century French mathematician Marie-Jean-Antoine-Nicolas de Caritat, Marquis de Condorcet. The hiring or admissions process can be viewed as a series of head-to-head contests between applicants. In the paradigmatic system, the candidate who is more qualified will prevail against another applicant for a particular slot. For example, between two applicants for law school in a perfect world,* the qualified candidate is the one who would become the better lawyer.

A generic player in this contest is designated $P$. A player who is successful in a contest will be labeled $SP$; the loser will be designated $LP$. Next, one can either be a qualified player $QP$ or an unqualified player $UP$. Using our previous idealized definition of *qualified,* we see that qualified and successful are not synonymous. Both qualified and unqualified players can succeed; we can have both a successful qualified player $SQP$ and a successful unqualified player $SUP$.

If we have both White players $W$ and African American players $A$, we can label them further as either qualified or unqualified: Qualified White $QW$, Qualified African American $QA$, Unqualified White $UW$, and Unqualified African American $UA$. Any of these players may succeed or lose in a given contest. We will label the successful as $SQW$, $SQA$, $SUW$, and $SUA$, while the losers will be labeled $LQW$, $LQA$, $LUW$, and $LUA$.

There are several ways to consider these categories. For instance, all the successful players can be counted by the equation $SP = SQW + SQA + SUW + SUA$. Similarly, the number of whites who lose the contest is found by adding together the losing qualified whites and losing unqualified whites: $LW = LQW + LUW$.

Now, let us turn to the debate. If affirmative action has *any* effect, it increases the number of successful African Americans. If this is a zero-sum game, the number of successful whites has decreased. Thus, there may well be tangible evidence that $SA$ is up

*Assuming there were still lawyers.

and $SW$ is down. But that is not what the debate is about. The question is whether the rate of qualified players becoming successful has improved. Mathematically, the issue is whether the proportion

$$\frac{SQP}{QP} \text{ is increasing, while } \frac{SUP}{UP} \text{ is decreasing.}$$

To create a mathematical portrait of the affirmative action debate, we must divide the above proportions further, into analogous proportions for African American and white players. We can describe the success rate for qualified players of each race as

$$\frac{SQW}{QW} \text{ and } \frac{SQA}{QA}$$

and for those who are unqualified yet successful as

$$\frac{SUW}{UW} \text{ and } \frac{SUA}{UA}$$

The success rate of all whites [or all African Americans] regardless of qualification is found by combining terms:

$$\frac{SQW + SUW}{QW + UW} = \frac{SW}{W}$$

$$\frac{SQA + SUA}{QA + UA} = \frac{SA}{A}$$

It can be seen that successful whites consist of both qualified and unqualified whites, $SW = SQW + SUW$, and that the same can be said for successful African Americans, $SA = SQA + SUA$. Similarly, we can see that the ultimate goal of the selection process, the successful qualified, consists of players of both races: $SQP = SQW + SQA$. Similarly, the error of the selection process, those who were successful although unqualified, is also populated by both races: $SUP = SUW + SUA$. Finally, we can see that the ranks of successful applicants contain both qualified and unqualified members: $SP = SQP + SUP$.

The problem with quotas, whereby a fixed proportion of success slots are reserved for one race, is that they directly affect the wrong category. The number of successful African Americans, *SA,* is indeed increased, but without any guarantee that it is only those who are qualified, *SQA,* who are being aided. Not surprisingly, quotas have long been declared unconstitutional by the Supreme Court.

+ + +

In *University of California v. Bakke,* Justice Lewis Powell wrote that "[R]ace or ethnic background may be deemed a 'plus' in a particular applicant's file, yet it does not insulate the individual from comparison with all other candidates for the available seats." Following this opinion, many institutions attempted to use race as merely one factor in hiring and admissions decisions.

The *Bakke* decision represented an attempt to focus the benefit on the qualified, without unduly benefiting the unqualified. But it did not quiet the debate. Opponents of affirmative action saw the continuing programs as helping "candidates who are *less qualified* than other applicants." Note that under my definition, there is no such concept as less qualified; either you are or are not the best person among those competing for that slot. Using my definition, this argument can be restated as a contention that unqualified players are still benefiting. In other words, the argument goes, there are more successful unqualified people, *SUP,* because there are more successful although unqualified African American candidates, *SUA*.

If that were the case, then more qualified parties would become unsuccessful. The innocent victim of affirmative action is the qualified white who loses a contest, *LQW*. In 1986, the Supreme Court ruled that it was unconstitutional to fire white teachers who had greater seniority than African American teachers who were retained: "[A]s the basis for imposing discriminatory *legal* remedies that work against innocent people, societal discrimination is insufficient and over expansive."

Proponents of affirmative action also focus their concern on those who are successful though unqualified, yet they draw attention to the unqualified white who wins, *SUW*. The innocent victim here is the qualified African American who loses a contest, *LQA*. According to this viewpoint, affirmative action is needed because the greater success rates for whites as opposed to African Americans,

$$\frac{SW}{W} > \frac{SA}{A}$$

can be attributed to an excessive number of successful unqualified whites, *SUW,* prevailing due to overt and covert racism, as well as to the lingering economic and social effects of discrimination: "You have but to ask any African-American man or woman, from the most accomplished to the least, to hear tale after heartbreaking tale of racially motivated unfairness, hostility, or even violence."

Under the earlier requirement of nonjingoistic, people of good faith should agree that it is preferable to minimize, if not eliminate, the number of innocent victims of all races, both *LQW* and *LQA*. As a society, we want the highest possible percentage, $SQP/QP \times 100$, of qualified players to be successful.

Of course, that requires maximizing the particular success rates for both qualified whites and qualified African Americans:

$$\frac{SQW}{QW} \text{ and } \frac{SQA}{QA}$$

But that cannot be done simply by equalizing either the number of successful whites, *SW,* and the number of successful African Americans, *SA*, or the percentage of successful members of each race as compared to their local population,

$$\frac{SW}{W} \text{ and } \frac{SA}{A}$$

As stated by the Supreme Court, it would be a " 'completely unrealistic' assumption that minorities [or majorities] will choose a particular trade in lockstep proportion to their representation in the local population." There are a host of nondiscriminatory reasons why one ethnic group, for example, may gravitate to one profession instead of another.

But what if there is a large difference between the success rates among members of both races who choose a particular career or educational path? Sometimes unspoken, though sometimes blatant, is the contention that such a large disparity might well be the natural order of things and that affirmative action is only needed, because, due to "chronic and apparently immutable handicaps, minorities cannot compete. . . ." It would be impossible, I would argue, to have a good faith discussion with someone who holds such a world view; it utterly fails the principle of nonracism, which rejects the noxious contention that one race is naturally better.

In contrast, there can be good faith disagreement in determining the degree to which a racial disparity is caused by either overt or subconscious discrimination. We only can estimate the degree to which the percentage of qualified African American applicants who succeed is lowered by illegal yet unprosecuted discrimination. We are similarly limited in ascertaining how much enforcement of existing antidiscrimination laws can smoke out all or most of these subtle and subconscious discriminatory practices.

To decide the mathematical appropriateness of affirmative action, we also must determine whether the decision-making system being used accurately compares differing candidates. To what extent can an admissions program rely on standardized scores as a proxy for quality and still adequately account for the rich variety of relevant factors? We know that all models are imperfect. Moreover, even in the best model, close questions can be excruciatingly difficult. One writer asks, for example, which candidate should a college accept: "the son of an African-American schoolteacher from Oakland, who eludes gangs and poverty, manages to earn a 3.2 grade point average in college prep courses and scores 1100 on the SAT . . . [or] the son of white Beverly Hills corporate lawyers who earns a 3.9 G.P.A., enrolls in an SAT prep course, and scores 1400 on the SAT?"

Race, however, is also an imperfect proxy for disadvantaged. Not all members of a given race face equal discrimination, nor are they equally qualified. Equating race with discrimination creates the same sort of problem as other uses of imprecise proxies. And, as before, the more flawed the proxy, the greater the harm to innocent victims.

Proponents of affirmative action respond that the number of innocent victims can be reduced by appropriate and careful use of race as merely one factor to consider. Moreover, they argue, especially in a nonmeritorious world, where children of alumni are assured a place in the winner's circle, race is a necessary albeit imperfect proxy to offset generations of privilege.

≠ ≠ ≠

In a fundamental sense, the ultimate problem raised by the debate about affirmative action is that neither side has a perfect model. Nor do I. One of the benefits of a mathematical approach is not that it will neatly resolve difficult questions but that it infuses analysis with an awareness of the inevitable imperfections in one's own position.

Imperfection is a particularly valuable concept. It allows people to concede that problems exist with their own positions; this then enables them to analyze both sides of the issue with less arrogant attitudes. Imagine a discussion on affirmative action beginning with the following points of agreement:

1. It is possible that a person who disagrees with me is acting in good faith.
2. The selection system I support represents an imperfect model. It does not create a pure meritocracy but uses various criteria to approximate merit.
3. Due to any system's imperfections, some qualified candidates, African American and white, are bound to be unsuccessful; moreover, some unqualified applicants will prevail regardless of the system used.
4. Not all numerical disparities indicate discrimination. But significant numerical disparities, coupled with our understanding of the history and current existence of discrimination, may indicate a problem.
5. Quantifiable factors such as test scores are imperfect measures for the important subjective and unquantifiable attributes, such as most qualified.

6. Race is an imperfect model for identifying those who are disadvantaged or victims of discrimination.
7. The weaknesses of any modeling system, even one created in good faith, create innocent victims.
8. A race-neutral mechanism does not guarantee that qualified applicants of all races will be identified. A race-based mechanism does not guarantee that only qualified applicants will be benefited.
9. Increasing our sophistication in defining qualified in an ideal sense and improving the proxies we use to approximate it are essential for avoiding the problem of the innocent victim.
10. Good faith will also be revealed by the energy people are willing to expend on finding acceptable solutions.

There will remain strong differences over the degree to which the various proxies are imperfect, as well as over the morality or immorality of using race as a factor. Nonetheless, a recognition that *all* systems are imperfect will, at least, remove some self-righteousness from the debate. Moreover, acknowledgment that the problem of innocent victims affects both sides may encourage a more concerted effort to find qualified people whoever they may be.

6

# Game Theory and the Constitution

**Interviewer: "One more question. Do you have any special advice for the young people who drive?"**
**James Dean: "Take it easy driving. The life you save might be mine."**

Game theory is a formal mathematical field that studies the behavior of people in situations where each person's outcome is influenced by the actions of other people. It is the rigorous mathematical study of "conflict between thoughtful and potentially deceitful opponents." Game theory is not about how to win a game such as chess or ice hockey. Rather, it presents ways to think about dealing with a contest in which each party's moves affect the others.

In game theory, a strategy is "a complete description of a particular way to play a game. . . ." Your best strategy (also known as your optimum strategy) must, by definition, be dependent on what your opponent is most likely to do. Thus, each player's best strategy reflects his or her opponent's best strategy.

As with other mathematical and economic models, game theory requires certain assumptions, which do not necessarily mirror the real world. The players are presumed to be perfectly logical. They also are presumed to be perfectly aware of the range of permissible moves for themselves and their opponents, as well as the ramifications of those moves. Additionally, the motivation of the players is assumed to be purely self-interested; they make "their decisions in an essentially amoral, self-serving egotistical fashion."

Leaving aside the contentious question of the effects these assumptions have on the capacity of game theory to predict and advise real people, the use of game theory can help to clarify issues that otherwise might be obscure. Game theory can provide valuable insight, even if it does not result in a complete explanation of situations where people's fates are intertwined.

## Constitutional Chicken

Game theory can illuminate many constitutional issues. Many constitutional crises, for example, can be seen as instances of jurisprudential games of chicken.

Chicken is one of a series of two-person, two-strategy games. (The best known of these games is the prisoner's dilemma, which is discussed later in this chapter.) The game of chicken was made famous in the 1955 James Dean movie *Rebel Without a Cause,* but it has been retold cinematically countless times. Depending on the story, two drivers are either heading off a cliff or straight toward each other. The first one to swerve out of harm's way is the chicken—the loser. Each would prefer to drive straight if the other swerves. If neither swerves, of course, they both crash and die, clearly the worst outcome for each. If they both swerve, they both live (which is better than death) but neither achieves the ego satisfaction of victory.

Numbering each party's outcomes on a scale of 1 (most preferred) to 4 (least preferred), a game theoretic model of chicken would look like this:

| **Other Guy** | **James Dean** | |
|---|---|---|
| | *Swerve* | *Drive Straight* |
| *Swerve* ➤ | 2, 2 | 3, 1 |
| *Drive Straight* ➤ | 1, 3 | 4, 4 |

The fundamental point of game theory is that even if an outcome is strongly desired, no player can select it unilaterally. The only power James Dean has is to select between the columns labeled "Swerve" or "Drive Straight" and accept the outcome which is produced with the "Other Guy's" choice of row. Accordingly, to determine one's best strategy for a particular game, the opponent's expected strategy must be taken into account.

In 1928, mathematician John von Neumann proved that for every two-person, zero-sum game, there is an optimal strategy. A strategy is optimal when neither player can improve his or her position by unilaterally choosing a different strategy. The output of these optimal strategies is known as an equilibrium point. In the 1950s, John Nash (who later won the Nobel Prize for economics) extended this theorem, proving that there was at least one equilibrium set of strategies in any multiplayer, nonzero-sum, noncooperative game. Modern game theorists refer to a set of equilibrium strategies as a Nash equilibrium.

In the game of chicken, there are two Nash equilibria: the two outcomes in which one player swerves and the other doesn't. Each represents an equally rational outcome, and there is no way to decide which path will be taken by rational players.

One party can gain the advantage in chicken, however, by demonstrating a credible commitment to driving straight. If you are convinced that your opponent is definitely or even probably going to drive straight, your only rational strategy is to swerve. You would then be chicken but at least not dead meat.

Thus, the player who makes a commitment to the more dangerous option, either by deed or simply by reputation, will defeat the rational player. But this is not a risk-free strategy. If both players make irrevocable commitments to drive straight, the resulting carnage produces the worst outcome for each.

× × ×

The first game of constitutional chicken involved the landmark case of *Marbury v. Madison*. The seeds of battle between President Thomas Jefferson and Supreme Court Chief Justice John Marshall began in November 1800, when Jefferson defeated John Adams for the presidency, and Jefferson's Republican Party captured control of Congress from the Adams-led Federalists. In the period between their electoral defeat and March 4, 1801, the start of the new administration, the Federalists attempted to secure control of the one remaining branch of the government—the judiciary. Secretary of State Marshall was confirmed by the Senate as chief justice on January 27 but continued as secretary of state until the end of the Adams presidency. Next, the Federalist Congress passed two laws: on February 13 a circuit court law, authorizing 16 new circuit judges, and on February 27 the Organic Act for the District of Columbia, creating 42 new justices of the peace. Adams was able to appoint and obtain Senate confirmation for these judges, but the paperwork for some of the justices of the peace could not be completed before March 3. President Jefferson ordered his secretary of state, James Madison, not to deliver their commissions to them. When one of those justices of the peace, William Marbury, brought suit in the Supreme Court to obtain his commission, the stage was set for a game of constitutional chicken between the Supreme Court and the president.

Faced with Marbury's request for a writ of mandamus—an order that Jefferson hand over the commission—Marshall had two choices analogous to those in the game of chicken: head for a crash by issuing the writ against the president or avoid the collision by not issuing the writ. Jefferson also had two choices: confront the chief justice by not delivering the commission or passively cooperate by handing it over.

Each player would benefit most from confronting a cowardly cooperator. Marshall would have gained the most from issuing an order and having the president comply. Jefferson obviously would have preferred to have had Marshall not issue the writ, so as to be able to deny the judgeship to Marbury. While a face-saving compromise, with Jefferson issuing the commission without a court order, would have been less desirable for each, the worst possible situation for each would have been for Marshall to have issued the writ and Jefferson to have refused to comply. Direct defiance of the order would have created a crisis of escalating proportions, leading either to a seriously weakened Court that could not enforce its order or a President condemned as a lawbreaker.

A game theoretic model of this situation might look like this:

| **Jefferson** | **Marshall** | |
|---|---|---|
| | *Not Issue Writ* | *Issue Writ* |
| *Give Commission* ➤ | 2, 2 | 3, 1 |
| *Not Give Commission* ➤ | 1, 3 | 4, 4 |

As with chicken, there is no obvious right move. However, if one party could create a credible threat that he is committed to the dangerous path, the other party's only rational move would be to avoid the subsequent carnage. Jefferson ensured the credibility of his threat not to comply with the Court's order by his refusal even to participate in the Court's hearing on the merits. On February 9, 1803, after closing arguments by Charles Lee, Marbury's attorney (and former attorney general under President Adams), no one rose to present Jefferson's argument: "It was obvious to all that Jefferson had deliberately intended to insult Marshall and his Court by cavalierly ignoring their authority."

Meanwhile, the Republican-controlled Congress was flexing its muscles. First, it repealed the Judiciary Act of 1801, removing from office the 16 Federalist judges. Moreover, it abolished the Supreme Court's June and December terms for 1802, so that the Court would not even be in session for more than a year, until February 1803. Finally, impeachment proceedings were brought against other Federalist judges. In February 1803, the House voted to impeach a Federalist district judge, John Pickering, and the Senate voted to convict him in March of that year.

In the midst of these machinations, Chief Justice Marshall could see that not only was President Jefferson committed to the destructive strategy of refusing to issue the commission but Congress was prepared to use its impeachment power to ensure that only the President would survive the "crash" of constitutional chicken.

With an opponent committed to refusing to cooperate, Marshall had only one rational course of action—to decline to issue a writ. But in a move of strategic brilliance, he did not base his inaction on the Court's inability to ensure compliance. Instead, Chief Justice Marshall justified not issuing the writ by writing an opinion striking down the law under which Marbury had brought suit. In so doing, he established the Court's power to strike down acts of Congress as violative of the Constitution. The effect of this move was that not only did Marshall avoid creating a crisis he knew he would lose but he empowered himself for future battles. The beauty of this move was that even though it strengthened the Court at the expense of the other two branches, the Republican members of those two branches supported the short-term outcome of not issuing the writ, and hence were unable to oppose the long-term effects of Marshall's strategy.

## The Prisoner's Dilemma and the Dormant Commerce Clause

In addition to describing such face-to-face conflicts, game theory also can illuminate certain confrontations that are likely to arise on the basis of the very structure of the Constitution. One such area is the tension between individual states competing for their share of the national economic pie.

The framers did not believe that altruism would govern commercial relations. When it comes to business, wrote Hamilton, "men are ambitious, vindictive, and rapacious." Legislators, who represent the interests of their constituents, would be expected to be inclined similarly. Democratic principles simply require that the ambitiousness, vindictiveness, and rapaciousness of the legislators be at the behest of the majority of the citizens.

But this democratic response to the desires of the local citizenry can lead to results that conflict directly with "the principle of the unitary national market. . . ." That is because one state may find it in the interest of its residents to harm their competitors in a neighboring state. Hamilton warned, "Each State, or separate confederacy, would pursue a system of commercial policy peculiar to itself. This would occasion distinctions, preferences, and exclusions, which would beget discontent."

To prevent such a divisive result, the Supreme Court has interpreted the Commerce Clause to include an implied Dormant

Commerce Clause. Under the Dormant Commerce Clause, states may not, in their dealings with one another, engage in protectionism—those actions "whose purpose or effect is to gain for those within the state an advantage at the expense of those without, or to burden those out of the state without any corresponding advantage to those within. . . ." Some might not mind so much if the federal government were to act this way against a foreign country, but there is no foreign entity among the several states. They are supposed to pass laws for their own benefit but not harmful to the national interest. Some modern commentators have observed that the issues underlying the Dormant Commerce Clause bear a strong resemblance to features of the game known as the prisoner's dilemma.

= = =

The prisoner's dilemma, like the game of chicken, is one of those game theory scenarios in which two players can choose whether to cooperate with one another or to defect. The difference between the two games is that the worst outcome for a player in chicken occurs if neither cooperates, while the worst outcome in the prisoner's dilemma occurs for a player who cooperates when his or her opponent does not. What makes the prisoner's dilemma the mainstay of much of game theory analysis is that the resulting optimum strategy leads to a conflict between apparent and true self-interest, between individual rationality and collective reasoning.

The basic story is this: Two prisoners, Ann and Bob, have been arrested and are being questioned by police. Ann is told that if she gives information to the police and Bob remains silent, she will walk away free, while Bob will go to jail for 10 years. If both speak out against the other (defection), they each will receive a moderately harsh sentence of 6 years. If neither speaks (cooperation), they can be convicted only of a lesser offense, and each will receive a light sentence of just 1 year. Bob is given a similar choice. Their options can be displayed as follows:

| **Ann** | **Bob** | |
|---|---|---|
| | *Cooperate* | *Defect* |
| *Cooperate* ➤ | 1, 1 | 10, 0 |
| *Defect* ➤ | 0, 10 | 6, 6 |

What is the best strategy for each? Ann rationally could think that if Bob cooperates and stays silent, she will receive 1 year in jail by cooperating with him but will be free if she defects and talks to the police. Defection then would be in her self-interest. If, alternatively, Bob defects and talks to the police, Ann would face 10 years in jail for cooperating with Bob but the lighter 6-year sentence if she defects and talks. Defection would be in her self-interest here, too. Thus, Ann could reason that whatever Bob does, her self-interest is to defect.

This sounds good, until you realize that Bob, acting equally rationally, will draw the same conclusion about his own self-interest. So Bob, too, will choose to defect.

The result of the mutual defection is the bottom right corner of our matrix, 6 years in jail for each. Of course, they each would have preferred only 1 year in jail, which they could have received by cooperating. Thus, by acting in what they rationally perceived to be their own self-interest, they not only hurt the other but themselves as well.

The only way out of the prisoner's dilemma is to change the rules to permit some enforcement mechanism. In repeated versions of the game, retaliation may be such a mechanism ("If you defect this time, I will defect next time"). Alternatively, two players can agree on an outside enforcement mechanism, some device that is outside the game ("If you defect, my brother will beat you up").

≥ ≥ ≥

Interstate trade conflicts possess many features common to the prisoner's dilemma. Sometimes cooperation is clearly in everyone's interest, as when Maryland and Virginia work together to protect the Chesapeake Bay. Other times, however, there may be economic motives for a state to help itself by hurting its fellow states, as when New Jersey prevented residents of Pennsylvania from using New Jersey's private landfills. In such circumstances, each state could think that mutual cooperation with free trade policies would provide some benefit but protecting their own residents when other states cooperated with free trade would lead to even greater economic advantage. Of course, if both states act to protect themselves, they each ultimately are placed in a much worse position.

We can create a trade matrix similar to that for the prisoners. Imagine two states, Texas and Oklahoma. Here, *to cooperate* means to adopt free trade policies and *to defect* means to engage in protectionism. I have switched the payoff values to reflect that in this version, each player wants *more* money, while each prisoner wanted *less* prison time.

| **Oklahoma** | **Texas** | |
|---|---|---|
| | *Cooperate* | *Defect* |
| *Cooperate* ➤ | 6, 6 | 0, 10 |
| *Defect* ➤ | 10, 0 | 1, 1 |

Texas will make a similar calculation to the one that Ann the prisoner did. Cooperation is nice, but for each choice of the other player, defection (meaning protectionism) creates a higher payoff. Oklahoma will reason similarly. Thus, each state will tend to choose protectionism (lower right corner), even though that reduces the economic well-being they would have enjoyed by cooperation (top left corner).

One way to prevent protectionism is retaliation. In modern terms, we would call this undesirable outcome a trade war. Another means is federal intervention. Congress has the power "to regulate Commerce . . . among the several states" and can preempt laws that interfere with interstate commerce. But that remedy often will not be forthcoming. The Supreme Court has noted that congressional inertia will permit many protectionist state laws to stand: "[T]hese restraints are individually too petty, too diversified, and too local to get the attention of a Congress hard pressed with more urgent matters." Thus, it has fallen to the Court to disentangle those actions of a particular state which are justifiable acts of sovereignties legitimately trying to serve the interests of their members from those which would cause unjustified disruption of our national economic unity.

Sometimes it is easy for the Court to determine that a law is improperly designed to bolster one state's economy by injuring those of other states. The clearest example is a protective tariff, which taxes goods imported from other states but does not tax similar items produced within the state.

To evaluate subtler forms of regulation, the Court has asked whether a law is "basically a protectionist measure, or whether it can fairly be viewed as a law directed to legitimate local concerns, with effects upon interstate commerce that are only incidental." In making this determination, the Court essentially is trying to determine whether a state would have passed the same legislation if it considered itself part of a unified economic whole rather than as a distinct economic unit, separate from the rest of the Union.

÷ ÷ ÷

The key question for the Court when it evaluates a Dormant Commerce Clause challenge has become identifying where the burdens imposed by a state's legislation fall. If burdens fall inside a state, that should not be a Dormant Commerce Clause concern. Intuitively, a state is not likely to pass a law where the harms outweigh the benefits to its residents. Voters would oust the legislators who inflicted the harm at the earliest opportunity. But, according to the Court, when a law's "burden falls principally upon those without the state, legislative action is not likely to be subjected to those political restraints which are normally exerted on legislation where it affects adversely some interests within the State."

Imagine that Minnesota is considering a law requiring that milk be sold in paper, not plastic containers. For simplicity's sake, assume that there are 40 people (such as plastics manufacturers or milk retailers) harmed by this law, who suffer a loss of $10 apiece, and that there are 20 people (such as paper manufacturers) helped by the law, who gain $10 apiece. If these people all reside in Minnesota, a majority would oppose the law and it would not be passed. But imagine that all 40 harmed by the bill are from other states. Minnesota legislators, responding to the wishes of the 20 in-state residents who will benefit, now will enact the legislation. Other states are worse off for this law by $400 and the country as a whole is worse off by the $200, but Minnesota, at least in the short run, has improved its situation by that same $200.

Thus, the Court has noted that "the existence of major in-state interests adversely affected . . . is a powerful safeguard against legislative abuse." When the burden of a law falls only on out-of-staters, there is no such safeguard. Because normal political restraint would not stop such a bill, the Supreme Court steps in under the Dormant Commerce Clause.

In *West Lynn Creamery, Inc. v. Healy,* the Court struck down a Massachusetts tax on milk sales that was coupled with a subsidy that effectively reimbursed Massachusetts milk producers. The Court held that the subsidy impermissibly skewed the political dynamics: "[O]ne would ordinarily expect at least three groups to lobby against the [tax]: dairy farmers, milk dealers and consumers." But, because of the subsidy, "Massachusetts dairy farmers, instead of exerting their influence against the tax, were in fact its primary supporters. . . ."

+ + +

Not all laws which have widespread local support and which place the primary burden on out-of-staters are bad. It is possible that certain laws are beneficial enough that they would have been enacted regardless of where the harm was felt. Consider this revised situation. This time 50 people are helped by the law and only 10 are hurt (again by $10 each). Now, in this case, whether the 10 victims are in state or out of state, the value of the law is such that the state is likely to enact it. The law provides $500 worth of benefit, and whether the $100 worth of harm is imposed in state or out of state, it makes economic sense to enact it. Thus, it should not be unconstitutional for a state to pass such a law, even if the victims happen to be out-of-staters.

A Maryland law banning companies that produced or refined gasoline from operating retail gasoline stations was upheld by the Court, even though 98 percent of the stations affected were owned by out-of-staters. The Court stated that "The fact that the burden of a state regulation falls on some interstate companies does not, by itself, establish a claim of discrimination against interstate commerce." That is because this law was seen by the Court as a reasonable attempt to protect gasoline consumers, which only incidentally harmed out-of-staters.

Sometimes, the Court has gotten itself into unnecessary difficulty in its struggle to distinguish those facially nondiscriminatory state regulations which are permissible from those which are impermissible. Occasionally, the Court has attempted to approximate the role of an idealized nonprotectionist legislator by explicitly trying to balance the costs and benefits of a law: "Where the statute regulates even-handedly . . . it will be upheld unless the burden imposed on such commerce is clearly excessive in relation to the putative local benefits." The basic theory here is that all commercial regulation produces winners and losers and a rational nonprotectionist legislator would not be willing to inflict harm that was "clearly excessive" in relation to the expected gain.

However, there are several problems inherent in this approach. First, there is no requirement in the Constitution that a law's benefit must outweigh the harm it causes. Legislators need not be rational, and legislation need not make sense overall. Moreover, unlike our earlier simplified examples, the total harms and benefits of legislation usually cannot be quantified readily. Thus, excessiveness is rarely determinable. This has led the Court to engage in such inappropriate judicial inquiries as whether the cost of Iowa's ban on long trucks, particularly for those truckers who had to bypass Iowa and drive around the state, exceeded the benefits derived by not having 65-foot trucks splashing passing Iowan drivers during bad weather. As Justice Antonin Scalia has complained, "I do not know what qualifies us to make the ultimate (and most ineffable) judgment as to whether, given the importance-level $x$ and effectiveness-level $y$, the worth of the statute is 'outweighed' by impact-on-commerce $z$."

≠ ≠ ≠

There are other, better methods for ferreting out protectionist legislation which avoid this neomathematical quandary. First, if a law directly discriminates against out-of-staters, it should be upheld only if it is "the least discriminatory alternative" that will accomplish the state's legitimate purposes. If the law is not facially discriminatory, it should be found unconstitutional only if its purpose is, "[s]hielding in-state industries from out-of-state competition."

While it is not always easy to identify an illegitimate purpose, it can be done. First, local politicians often will reveal their true protectionist motivation in order to receive due credit from in-state voters. In the case of the Iowa trucking ban, for example, the governor of Iowa stated that he opposed permitting long trucks, because it "would benefit only a few Iowa-based companies while providing a great advantage for out-of-state trucking firms and competitors at the expense of our Iowa citizens."

Even a more sophisticated protectionist state government, one that only expresses a legitimate purpose—such as safety or consumer protection—will be found out if, in the words of then-Associate Justice Rehnquist, the "asserted justification . . . is merely a pretext for discrimination against interstate commerce." Such a pretext has been discerned in a statute that harmed out-of-staters and did "remarkably little to further [the state's] laudable goal." For example, the Court struck down a North Carolina law that prohibited the sale of apples in crates marked with labels indicating their state grading system. The law, which North Carolina claimed to be a consumer protection law, would have harmed only those states with higher quality apples, such as Washington, by preventing the highlighting of their superior product. When the Court invalidated the law, it did not need to weigh its costs and benefits. Instead, the Court considered whether the suspicion of protectionism, which arises whenever a special burden is imposed on out-of-staters, was borne out by a finding that the actual benefit was "illusory, insubstantial, or non-existent."

Thus, the Dormant Commerce Clause illustrates both the strengths and limitations of applying mathematical thinking to the Constitution. Game theory indicates that state protectionism is not an unlikely occurrence in our federalist system. Mathematical analysis shows that majoritarian impulses may well cause states to pass laws that are detrimental to the economic interests of other states and the nation as a whole, as long as there are benefits for local concerns. But the Court should not rely on some mathematical analysis to decide whether a law's costs to out-of-staters outweighs its benefits to local residents. Identifying hidden protectionist legislation is ultimately a political and legal, not an arithmetic, determination.

# 7

# Multidimensional Thinking

**[T]he rational person who has grasped the lessons of non-Euclidean geometry is at least wary of snares, and, if he accepts any truths, he does so tentatively, expecting at any moment to be disillusioned.**

***—Morris Kline,***
**Mathematics in Western Culture *(1953)***

Robert Osserman, the deputy director of the Mathematical Sciences Research Institute, describes a pattern for the reception of new mathematical concepts that has parallels in most areas of intellectual endeavors—the path from ridicule to orthodoxy:

> *One theme that recurs time and again in the history of mathematics is the gradual evolution of a new concept—from its initial rejection as being too abstract, through grudging acceptance of its usefulness, despite the fact that it appears "unnatural" and counterintuitive, to its eventual elevation to the status of a basic and indispensable tool in applications.*

It took many centuries, for example, for the concept of a negative number to be accepted. If numbers were to be used to count or measure the real world, how could you have a negative distance or negative number of rocks? Of course, anyone who is careless with a credit card quickly learns just how easy it is to have a negative amount of money, and mathematicians, too, eventually realized the utility and importance of the concept.

Of all the citadels of mathematical certainty, none were held in higher regard than the axioms and theorems of Euclidean geometry (discussed in Chapter 2). Not only did the techniques of axiomatic logic appear untouchable, the fundamental axioms or postulates and the hundreds of theorems that were derived from them were seen as the perfect model to describe our world. The infallibility of Euclid was a mathematical "fighting faith."

Yet, in the 1800s, mathematicians began to suspect that there might be other ways of thinking about geometry. The first four of Euclid's five postulates seemed too unexceptional to question:

1. A straight line may be drawn between any two points.
2. Any straight line may be extended indefinitely.
3. A circle may be drawn with any given point as center and any given radius.
4. All right angles are equal.

Attention was focused on the last of Euclid's five postulates, known as the parallel postulate: In a plane, through any point not on a given line, there is one and only one line parallel to that given line. This postulate comports well with common sense. If you draw a straight line with a ruler and put a dot above it, it seems perfectly clear that there is only one line you can draw through that dot that will not touch the first line, no matter how far those lines are extended.

Nonetheless, unlike a theorem which can be proven from existing postulates, the parallel postulate had to be taken on faith. As with all postulates, it could be neither proven nor disproven from the other postulates. Thus, if one were able to describe a situation where there were no parallel lines or, conversely, an infinite number of parallel lines through a given point, a new, tradition-breaking geometry could be created.

The easiest non-Euclidean geometry to visualize was developed in 1854 by Georg Friedrich Riemann and is grasped most readily as the geometry on the surface of the sphere. Imagine that our Earth, like a typical globe, is a perfect sphere. If you take any two lines that are perpendicular to the equator (longitude lines), they will always meet at two points, the North and South Poles. Unlike on a flat plane, there are no parallel lines on a sphere.

Not only that, but once we start thinking of a sphere instead of a plane, we also must reconfigure what we mean by a line. If we maintain our usual understanding that a line is the shortest distance

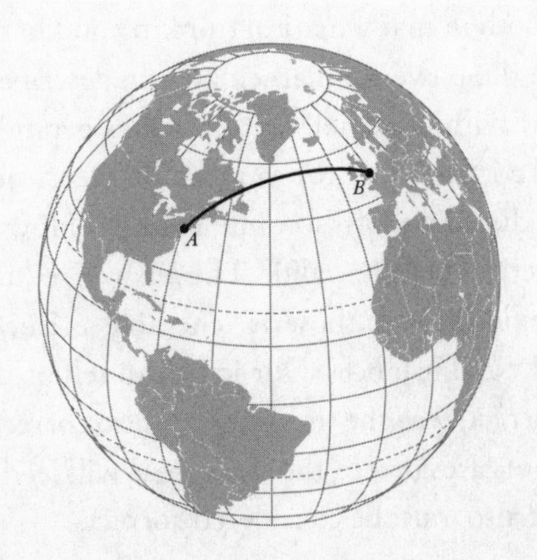

between two points, then a line on a sphere is what is termed a geodesic, which is an arc of a great circle. A great circle cuts the sphere into two equal hemispheres and, on our globe, has the same center as the equator. That's why the flight path between Baltimore and London does not look like a straight line on a flat map.

Moreover, many of the theorems of Euclidean geometry are altered in Riemann's non-Euclidean geometry. A triangle can have three right angles, for instance. Also, two points can determine an infinite number of lines (witness the North and South Poles).

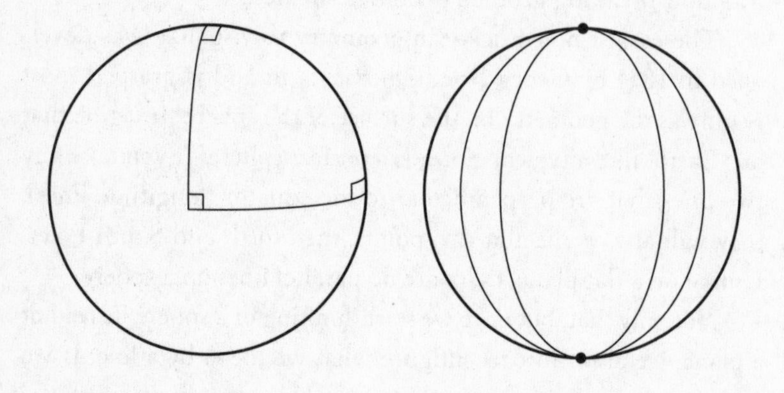

The eventual acceptance of non-Euclidean geometry created a revolution both in mathematical thinking and in the confidence people had in the power of mathematics to describe our world. If you measured within a small area, Euclid's geometry of the flat surface worked fine. But for very large areas, non-Euclidean geometry produced a more accurate reading. How could contradictory geometries both be valid? The answer required accepting the fact that axioms, even those as venerable as Euclid's, were not correct for all circumstances. Our logic can tell us that *if* Euclid's axioms are correct, *then* the theorems are also correct. However, if different axioms are correct, the same logic will lead to alternative theorems that also must be considered correct.

## Non-Euclidean Imagination and the Constitution

One mathematician proposed that the difference between Euclidean and non-Euclidean geometries has an analogue that can be found by comparing the Declaration of Independence with the Gettysburg Address. Lipman Bers noted that the Declaration of Independence had declared it a truth that all men were created equal, while the Gettysburg Address merely stated that the nation was "dedicated to the proposition that all men are created equal." Thus, while Jefferson, in the spirit of Euclid, announced his axiom of equality, "Lincoln's frank unwillingness to commit himself to the propositions he was citing" mirrored the philosophy of non-Euclidean geometry.

There is much that students of the Constitution can derive from non-Euclidean geometry. For starters, it provides a valuable lesson in humility. We can realize that others can hold axioms, or core principles, that contradict our own, yet contain their own equally valid logic.

On the broadest level, we must understand that our Constitution is not the only valid one. England has a parliamentary government and France has a more limited form of judicial review, yet democracy flourishes in both those countries. As in the comparison of Euclidean and non-Euclidean geometry, our different systems have many postulates of freedom in common, but a few are quite different, due to both historical and cultural reasons.

The study of other countries' constitutional systems offers great insights into our own. We can learn that what we take for granted may not seem so obvious to others. France grants its citizens a right of reply to disparaging newspaper articles, a concept that was unanimously rejected by the U.S. Supreme Court in *Miami Herald Publishing Co. v. Tornillo*. In England, courts do not question the binding authority of an act of Parliament. The

United States's system of judicial supremacy has endured since *Marbury v. Madison*.

We must appreciate that there is more than one possible constitutional geometry. But not all systems should be considered valid. The legal systems of Nazi Germany or Communist Russia produced geometries of horror. Even if the results flowed logically from each system's own constitutional postulates (of ethnic superiority or the omniscience of government), on some objective standard we would say that such social and political axioms must be rejected.

× × ×

The story of non-Euclidean geometry serves another important function: It warns us against overreliance on timeworn ways of thinking. As mathematical historian Morris Kline wrote, "Paradoxically, although the new geometries impugned man's ability to attain truths, they provide the best example of the power of the human mind, for the mind had to defy and overcome habit, intuition, and sense perceptions to produce these geometries."

An example of such intellectual growth can be seen in the metamorphosis of Thomas Jefferson's views on race and intelligence. In his *Notes on the State of Virginia*, Jefferson had explained his belief in the "real differences" between African Americans and whites: "[I]t appears to me, that in memory they are equal to whites; in reason much inferior, as I think one could scarcely be found capable of tracing and comprehending the investigations of Euclid. . . ."

In 1791, Benjamin Banneker, the son of former slaves, sent a manuscript of his almanac to Jefferson. This almanac was a work of extraordinary mathematical and scientific scope, and Banneker's cover letter to Jefferson argued that it proved that "one universal Father hath . . . endued us all with the same faculties." The almanac apparently had the desired effect, for within less than a week of receiving it, Jefferson wrote of Banneker in the following letter to the French mathematician Marie-Jean-Antoine-Nicolas de Caritat, Marquis de Condorcet:

> *I am happy to inform you that we have now in the United States a negro, the son of a black man born in Africa, and a black woman born in the United States, who is a very respectable mathematician. . . . I have seen very elegant solutions of Geometrical problems by him. Add to this that he is a very worthy & respectable member of society. He is a free man. I shall be delighted to see these instances of moral eminence so multiplied as to prove that the want of talents observed in them is merely the effect of their degraded condition, and not proceeding from any difference in the structure of the parts on which intellect depends.*

There is a noteworthy mathematical irony here. The mastery of Euclid was used by Jefferson to determine mental capacity objectively. But Jefferson's own ability to change his perceived understanding reflects the power of a non-Euclidean mind-set.

## What Shape Is Our Government In?

No one wants to be called one-dimensional. The term connotes a person with only one interest, one goal, or one perspective. To be considered multidimensional makes you sound more interesting. There is more than one side to you; you are more complex.

One constitutional lesson from the study of dimensions can be drawn from the relationship of length (the measurement of a single dimension) to area (the measurement for two dimensions) and finally to volume (the measurement for three). A particularly interesting facet of this relationship is that a small change in length can lead to a large change in area and a very large change in volume.

Let's imagine a box whose sides are 10 inches long. The area of each of the box's square faces is 100 square inches ($10 \times 10$), and the box's volume is 1,000 cubic inches (10 × 10 × 10). If we double the length of a side of the box, we've increased the length by 10 inches, from 10 to 20. The area of each face, though, has grown fourfold, from 100 to 400 square inches. Moreover, the volume of the box is 8 times larger, from 1,000 to 8,000 cubic inches. That is, when the length is doubled, the area is raised by a factor of 4, and the volume by a factor of 8.

What this means physically is that a structure that is designed to support one size cannot grow indefinitely. In fact, "whatever shape you design, if built on a large enough scale, will collapse under its own weight. An elephant-sized beetle would never get off the ground; its legs could not sustain the weight."

This concept explains why a local government, like a city, is not likely to have the same "shape" as the federal, or even its own state, government. Bicameralism (a two-house legislature), for example, which is found in nearly every state house, usually is not a feature of even the largest city government.

= = =

The structure of constitutional power sharing can be envisioned in many different mathematical ways. Using Madison's imagery of a "sphere of jurisdiction," we can consider the ways in which two different spheres can intersect: They can meet at no points, at only one point, at several points, or one can be subsumed completely by the other.

An area in which one sphere of jurisdiction is indeed subsumed by another (a) is the relationship of city governments to state governments. Under our constitutional system, "a municipality is merely a political subdivision of the State from which its authority derives." Any power exercised by the local government is deemed to be derived from a grant by its state.

For an example with no intersection (b), consider the case of

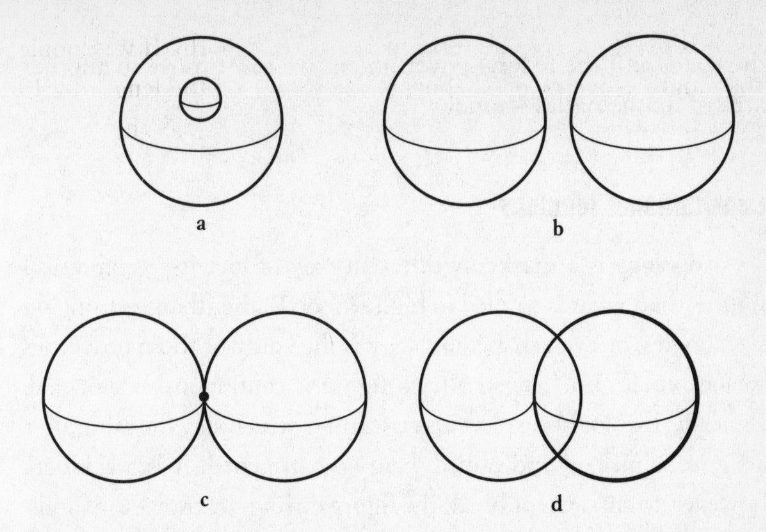


the president and the process of amending the Constitution. Article V gives absolutely no role to the chief executive in the proposing or approval of amendments to the Constitution.

It is more difficult to come up with two spheres with just a single point of intersection (c), but one candidate might be the role of the judiciary in impeachment proceedings. The House of Representatives has the "sole Power of Impeachment" and the Senate has the "sole Power to try all Impeachments." There is only one responsibility given to the judiciary in impeachments, and that occurs when the president is being tried by the Senate. For that trial alone, "the Chief Justice shall preside." During the impeachment of President Clinton, for example, the Senate trial was presided over by Chief Justice William Rehnquist, resplendent in his black robe with four gold stripes on each sleeve.

The intersections between the federal and state government spheres are among the most complicated in all of constitutional law. Numerous points of intersection exist (d). The states are neither subsumed by nor independent of the national government. To understand some of the subtleties of the relationship between the states and the federal government, we must move to another area of mathematics—topology.

## Constitutional Topology

Topology is a markedly different way of viewing geometrical figures than what is studied in high school. Rather than focusing on size, angles, or even shape, topology is the study of those properties which remain unchanged after a figure is continuously deformed. Basically, topologists explore those features which stay the same after a figure is pushed and pulled, bent and stretched. It is forbidden, however, to cut, tear, or break the figure during the transformation.

Topology is sometimes called rubber-sheet geometry, because it treats shapes as if they were drawn on a highly flexible surface. Suppose you press a piece of Silly Putty against a comic strip. The image you pick up looks normal, but what if you stretch and pull the Silly Putty? Shapes change and the cartoon characters become deformed, yet the words of dialogue, even if unreadable, stay inside the bubble in which they originally were drawn.

In old-fashioned Euclidean geometry, shapes were seen as rigid, unchangeable by motion. One could take a rectangle, and

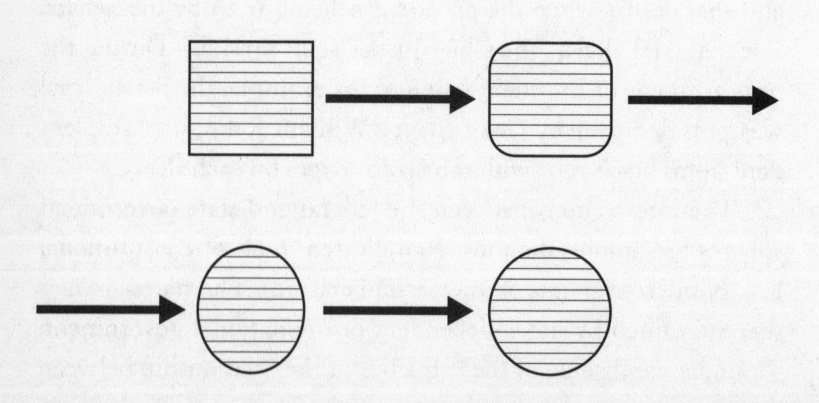

after rotations and reflections, it would be the same rectangle, with the same length sides and the same right angles. But if you envision a shape as flexible, capable of responding to pushing and pulling, you can see that if you stretch out that rectangle, it soon stops looking like a rectangle at all: In fact, you can turn it into a circle. Thus, the state of being a rectangle is not a topological property.

Topology looks for "deeper invariances." For example, the outer lines of the original rectangle stayed connected to one another, and there are no holes at either the beginning or the end. These traits are considered topological properties. Topologists, being a playful lot, enjoy transforming a coffee cup into a doughnut. Keeping in mind that all topological figures are flexible, we can push together the top rim and bottom surface of the cup, which removes the space that normally holds the coffee, and then stretch and push the handle into a doughnut hole.

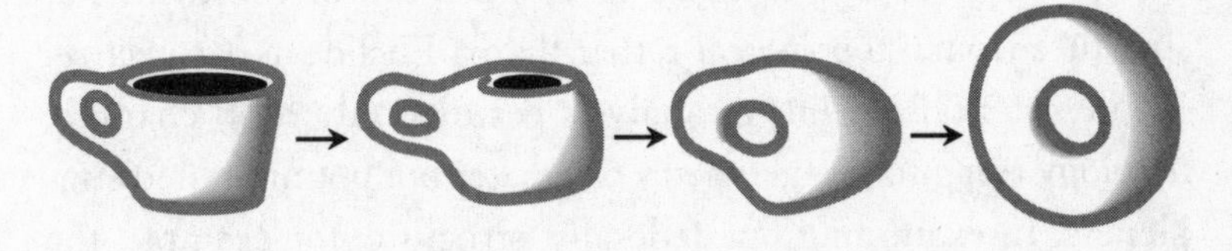

*Figure from Ian Stewart,* Concepts of Modern Mathematics *(1975), 46.*

The number of holes (or equivalently, handles) that an object has is a topological invariant of that object. Thus, while our coffee cup could be transformed into a doughnut, it cannot become a blueberry muffin, an object without a hole.

The number of holes a figure possesses is called its genus. One of the fundamental theorems of topology is, in simplified form, that every closed surface with some number of holes is topologically equivalent to a sphere with the same number of holes (the same genus). Thus, a torus (the mathematical name for a doughnut shape) can be transformed into a sphere with one handle, and a two-holed torus can be manipulated into a sphere with two handles. But the different toruses cannot be transformed into one another.

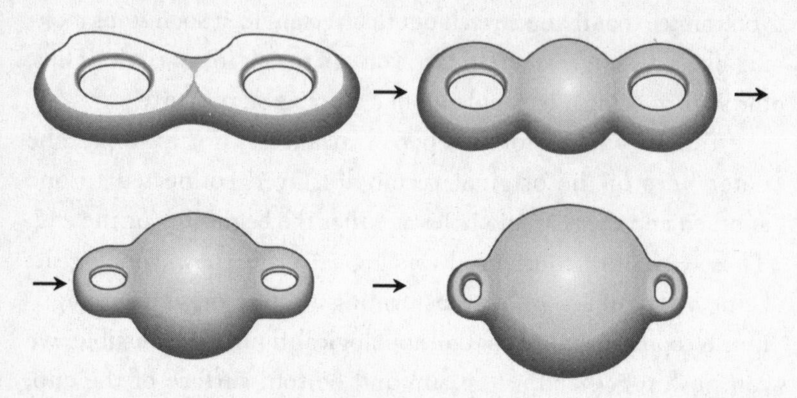

*Figure from Keith Devlin,* Mathematics: The Science of Patterns *(1997), 184.*

In many ways, it is more useful to think of constitutional structure from a topological rather than a Euclidean perspective. While classical Euclidean analysis permits only rigid changes, topology recognizes a geometry of greater but not unlimited suppleness. In examining our federalist structure, for example, the constitutional "framework has been sufficiently flexible over the past two centuries to allow for enormous changes in the national government." Yet, at the same time, "the power of the Federal government is subject to limits that . . . reserve power to the States. . . ." In other words, some changes can be seen as merely stretching or bending our system, but others would go so far as to tear apart our fundamental structure.

≥ ≥ ≥

The division of power between the national and state governments is, in the words of Justice Anthony Kennedy, "the unique contribution of the Framers to political science and political theory." Paradoxically, perhaps, the framers saw that liberty was better protected if each citizen were under two governments, rather than only one. As James Madison wrote in *The Federalist Papers*,

> *In the compound republic of America, the power surrendered by the people is first divided between two distinct governments, and then the portion allotted to each subdivided among distinct and separate departments. Hence a double security arises to the rights of the people. The different governments will control each other, at the same time that each will be controlled by itself.*

While the two governments were never meant to be completely distinct from one another, the relationship between the two spheres has changed enormously over time. For example, in 1895, the Court held that Congress had no power to prevent a nationwide monopoly over factories refining sugar, because manufacturing was purely a local concern. By 1942, however, the Court had expanded greatly the power of Congress to regulate interstate commerce. The Supreme Court upheld federal regulation of a small farmer's growing wheat for his own consumption, on the theory that his local activity, combined with that of others, would have a substantial impact on interstate commerce. This expansion of congressional power over commerce evolved both as a reaction to the Great Depression, as well as a recognition of the changing nature of the way in which business was being carried on in this country.

Yet, in 1995, the Court ruled that there were constitutional limits on federal power and, in particular, that Congress had no power to make it a federal offense to possess a gun near school property. The Court held that to permit Congress to regulate such noncommercial activity on the theory that there would be an eventual impact on interstate commerce would be to conclude "that there never will be a distinction between what is truly national and what is truly local. This we are unwilling to do."

Thus, while Congress can regulate local economic matters, such as minimum wages and antitrust laws, it is barred from regulating local criminal activity with no direct impact on commerce.

A related pattern can be seen in the Court's view of Congress's power to regulate states directly. Reversing a 1976 decision, the Court in 1985 upheld the power of Congress to impose minimum wage requirements on state governments. The Court rejected the argument that "the States as *States* have legitimate interests which the National Government is bound to respect. . . ." However, the Court subsequently has held that Congress can compel neither a state legislature to enact a particular regulatory scheme nor state officers to administer or enforce a federal regulatory program. The national government may not commandeer state governments, "reducing them to puppets of a ventriloquist Congress."

The evolving relationship between our federal and state governments can be examined topologically. Imagine the federal government as a sphere containing all of the powers granted by the Constitution. Within that sphere, however, there exists a hole, consisting of the powers that are, in the words of the Tenth Amendment, "reserved to the States." Over time, the size of the hole has grown and shrunk relative to the size of the sphere, but the hole must remain if the Constitution's topological structure is to remain intact. In preserving "our constitutional system of dual sovereignty," the Court has maintained the integrity of the constitutional framework. In this case, it is imperative to watch both the doughnut and the hole.

÷ ÷ ÷

Another topological exercise reveals the constitutional shortcomings of the legislative veto. In *INS v. Chadha*, the Supreme Court held that giving Congress the ability to override the decision of an executive agency was a violation of the separation of powers. The issue of *Chadha* can be illustrated through the mathematical problem that served as the foundation of topology, the seven bridges of Königsberg.

This Prussian city was divided in two by the Pregel River. Additionally, there were two small islands in the river. The four sections of the city were connected by seven bridges. The townspeople, including, allegedly, philosopher Immanuel Kant, strolled freely across these bridges. The question ultimately was raised whether it was possible to go on a round-trip stroll across all seven bridges without crossing any bridge more than once.

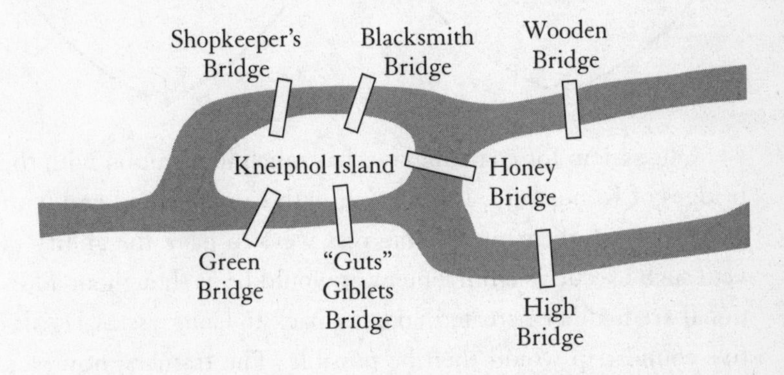


*Figure from Paul Hoffman,* Archimedes' Revenge *(1988), 161.*

In 1736, the brilliant mathematician Leonhard Euler proved that such a journey was impossible. He reduced the problem to a simple geometric picture, with the land masses as points and the bridges as arcs, and showed that the solution could be derived from calculating how many points (land masses) were connected to an odd number of arcs (bridges). He then was able to prove that, no matter how many bridges existed, a non-overlapping trip could be taken only when either every point was connected to an even number of arcs or exactly two points were connected to an odd number of arcs. This follows from the observation that one must be able to enter *and* leave every point except for the beginning and the end.

If we consider in the sketch below the simple path on the left, a nonrepeating round trip is impossible. But if an additional arc is drawn as on the right, the solution changes. Our walker can still visit all of the points, but this time he would finish where he started.

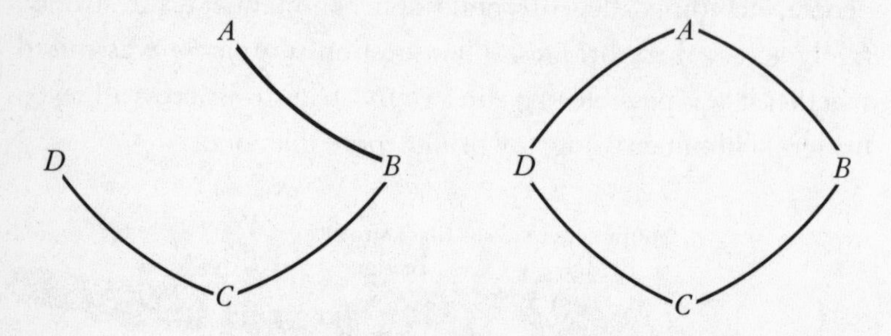


Our system for enacting laws has much in common with the bridges of Königsberg. Laws begin with Congress and end with executive enforcement. If Congress were to have the ability to veto such executive enforcement, it would be as though an additional arc had been created, looping back to Congress. A legislative round-trip would then be possible. The framers, however, chose to create a system in which the enforcement of our laws rests with the president, not Congress. Whether this is the best choice is not the question; it is enough to say, as the Court did in *Chadha*, that this is not the path laid out in the Constitution.

## Relativity

In 1905, Albert Einstein proved that two people theoretically could come up with different measurements for the distance an object traveled between the same two points, and the two different measurements both could be correct. To begin with, consider two people, George and Martha, standing on a flat surface, with

Martha standing 50 feet to the east of George. If George throws a ball to Martha, both of them will say that the ball traveled 50 feet.

However, imagine that the flat surface is the floor of a train moving at extraordinarily high speed in an easterly direction. If a third person, a mere observer, is standing on the platform watching the train speed by, the observer will have a very different sense of the distance the ball traveled. If the train (and of course Martha) have traveled 10 feet to the east from the time the ball was thrown to the moment it was caught, the observer will measure the ball as having traveled 60 feet.

George➤------------------------ Martha *Throwing Position*

George➤------------------------Martha *Catching Position*

0 50 60

Observer

So, did the ball travel 50 feet, as it appeared to George and Martha, or is the observer correct in claiming the ball traveled 60 feet? Einstein's answer was a remarkable "It depends." There is no way to say whether either viewpoint is correct. It is literally meaningless to describe the distance without first deciding whose perspective to use in measuring that distance.

Einstein went on to establish that not only was the concept of distance relative, so were such seemingly fixed concepts as velocity and time. They all depend on a preliminary determination of a point of reference. Not everything in the universe is relative, however. Most importantly, the speed of light is invariant, unchanging from any perspective.

+ + +

The importance of distinguishing between relativity and invariance can be seen in the infamous case of *Plessy v. Ferguson*. In upholding a Louisiana law requiring railroad companies to provide separate coaches for whites and African Americans, the Supreme Court declared that any perception that the law inflicted a "badge of inferiority" was only in the minds of those challenging the law: "If this be so, it is not by reason of anything found in the act, but solely because the colored race chooses to put that construction upon it."

Justice Harlan argued in lonely dissent, that the racial hostility and animus reflected in Jim Crow laws was not a matter of relative perception but a universally acknowledged invariant:

> *Every one knows that the statute in question had its origin in purpose, not so much to exclude white persons from railroad cars occupied by blacks, as to exclude colored people from coaches occupied by or assigned to white persons. . . . What can more certainly arouse race hate . . . than state enactments which, in fact, proceed on the ground that colored citizens are so inferior and degraded that they cannot be allowed to sit in public coaches occupied by white citizens? That, as all will admit, is the real meaning of such legislation as was enacted in Louisiana.*

≠ ≠ ≠

Deciding when and how the Court should consider relativistic points of reference can be very difficult, and nowhere is it more complex than in dealing with the issue of the establishment of religion. For example, the Court has ruled that the Establishment Clause, "at the very least, prohibits government from appearing to take a position on questions of religious belief or from 'making adherence to a religion relevant in any way to a person's standing in the political community.'" To determine whether the government "appears" to take a position on a religious question, the

Court decides whether the challenged government action, such as placing religious symbols on public property, "is sufficiently likely to be perceived by adherents of the controlling denominations as an endorsement, and by the nonadherents as a disapproval, of their individual religious choices."

This quest, however, is made far more complicated by the fact that the perceptions of adherents and nonadherents may diverge so widely as to be almost incomprehensible to the other. Determining how an action is perceived by both adherents and nonadherents demands extraordinary sensitivity. It requires, above all else, a willingness to understand that perceptions differ widely from one observer to the next. Religious beliefs are tied so intrinsically to an individual's cultural upbringing, sense of self, and view of the world that different religious beliefs will lead to vastly different perceptions.

The Court has had great difficulty reaching a consensus as to how to deal with these varying perceptions. For example, in *Lynch v. Donnelly*, the Court permitted a city to include a nativity scene in a display in a public park, which also contained hundreds of colored lights, a Santa Claus house, a clown, an elephant, a teddy bear, and candy-striped poles. Speaking for the Court, Chief Justice Warren Burger found no message of religious endorsement: "The display engenders a friendly community spirit of good will in keeping with the season." Justice Harry Blackmun, in dissent, argued that permitting governmental displays of the crèche if and only if it is surrounded by nonreligious symbols, is offensive to all: "Christians feel constrained in acknowledging its symbolic meaning and non-Christians feel alienated by its presence."

The perceptions of nonadherents have been central to the Court's rulings on prayer in public schools. Even when the children who object to a prayer are permitted to remain silent or leave the room, the Court has recognized that those children may be receiving a very different message from their classmates: "What to most believers may seem nothing more than a reasonable request that the nonbeliever respect their religious practices, in a school context may appear to the nonbeliever or dissenter to be an attempt to employ the machinery of the State to enforce a religious orthodoxy."

Consistent with the lessons of relativity theory, both these contradictory views can exist simultaneously, and their validity depends on the perspective of the observer. It falls to the Court to decide which perspective to adopt if the views cannot be reconciled. Because of the need to protect vulnerable children, the Court has chosen to rule from the position of the nonbeliever or dissenter, and thus bans even voluntary teacher-led prayer in public schools.

× × ×

There is an important misunderstanding that some people have about the theory of relativity, both as a mathematical concept and as a metaphor. The fact that observations are relative does not support the nihilistic conclusion that no knowledge is possible because everything is relative. Relativity is not an excuse to renounce the search for meaning and truth. Mathematically, for example, through equations known as Lorentz transformations, we can calculate precisely the time and distance associated with other people's perspectives. Thus, for any group of observers, we can compute which differing time and distance measurements can be valid. Analogously, in terms of our understanding of the Establishment Clause, the only perspectives that need be respected are those that are reasonable.

Consider the case of *Capitol Square Review and Advisory Board v. Pinette*, in which a sharply divided Court ruled that the month-long placement by the Ohio Ku Klux Klan of a large cross in a public park in front of the Ohio Statehouse did not violate the Establishment Clause. Justice Scalia, for a four-justice plurality, argued that private speech on a public forum, which is protected by the First Amendment, never can be viewed reasonably as governmental endorsement: "Private religious speech cannot be subject to veto by those who see favoritism where there is none."

Justice O'Connor agreed that the Klan's cross did not violate the Establishment Clause but said that, under other circumstances, private use of a public forum possibly could amount to a violation. She defined the "reasonable person" as "a hypothetical observer who is presumed to possess a certain level of information that all citizens might not share."

Justice Stevens, dissenting with Justice Ruth Bader Ginsburg, argued that the large cross in front of the statehouse would indeed communicate government approval to a reasonable observer and argued that not all reasonable people see things the same way: "It is especially important to take account of the perspective of a reasonable observer who may not share the particular religious belief it expresses."

Despite this widespread disagreement on defining the reasonable person, there are still some perspectives which none of the justices would consider valid. As Justice Stevens stated, "A person who views an exotic cow at the zoo as a symbol of the Government's approval of the Hindu religion" would not be objectively reasonable.

= = =

Knowing that not all perspectives are reasonable, however, does not identify the ones that are unreasonable. Sometimes, it seems, the courts have been all too willing to devalue certain religious perspectives. In the case of *Mozert v. Hawkins County Board of Education*, a federal court of appeals ruled that a public elementary school could require even deeply religious pupils to read aloud from a reading list which "denigrates and opposes their religion." According to the court, parents who were fundamentalist Christians wanted their children excused from reading several books that they claimed included many themes repugnant to their religious beliefs, including evolution and the occult. The appellate court rejected the parents' perception of the conflict between the readings and their religion: "While many of the passages deal with ethical issues, on the surface at least, *they appear to us to contain no religious or anti-religious messages*. Because the plaintiffs perceive every teaching that goes beyond the 'three R's' as inculcating religious ideas, they admit that any value-laden reading curriculum that did not affirm the truth of their beliefs would offend their religious convictions." Thus, according to one of the concurring judges, "[T]he school board is indeed entitled to say, 'my way or the highway.'"

The problem with this case can be seen by conducting what Einstein referred to as a thought experiment. Picture two high school seniors, Alice, an atheist, and Frank, a fundamentalist Christian, in a classroom. The teacher begins the day with a voluntary nonsectarian prayer. Next, the teacher administers a science test, which must be passed before a student is permitted to graduate. Several questions ask the age of various rocks and prehistoric fossils. Frank knows that the answer deemed correct by his teacher violates the teachings of his religion on the age of the universe.

Just as Alice has "a reasonable perception that she is being forced by the State" to comply "in a manner her conscience will not allow," so does Frank. For both, "the Constitution forbids the State to exact religious conformity from a student" as the price of attending a public high school. But Alice will prevail in her constitutional challenge to the teacher-led prayer, while Frank must choose between failing a course in a public school and forsaking his religious beliefs.

If the differing perspectives are not to be protected equally, we surely need a better rule than that when the school authorities disagree with nonfundamentalist students, the school loses, but when the school authorities disagree with fundamentalist students, the school wins. It could be argued that the difference lies in the legalistic distinction that Alice is claiming a violation of the Establishment Clause, and Frank is arguing free exercise. Perhaps it could be shown that excusing students from the classes or changing the exam questions "would result in substantial disruption to the public schools." But courts, at the barest minimum, should be highly reluctant to discount either student's perspective as unreasonable.

≥ ≥ ≥

The issue of sensitivity to the religious beliefs of others has an analogue in topology as well as relativity. One of the elementary theorems of topology is that "Every closed curve in the plane which does not cross itself divides the plane into one inside and one outside." Simply put, if you have a circle, square, or any squiggly closed shape, it will divide a surface into those points which are inside and those which are outside.

Much of the history of our Constitution can be categorized as an attempt to alter the topology that divides us. Today, dividing the populace between insiders and outsiders is rightfully considered distinctly un-American. Our law no longer accepts the divisive view that "this is a Christian nation" or that due to "racial instincts or . . . distinctions based on physical differences . . . one race [is] inferior to the other socially."

The geometrical solution to such antiquated notions may, in the final analysis, be elegantly simple. In the words of the poet, Edwin Markham,

*He drew a circle that shut me out—*
*Heretic, Rebel, a thing to flout.*
*But Love and I had the wit to win:*
*We drew a circle that took him in!*

# 8

# Infinity and the Constitution

**Just as sight recognizes darkness by the experience of not seeing, so imagination recognizes the infinite by not understanding it.**

***—Proclus (412–485 A.D.)***

## Hilbert's Hotel

There once was a very popular hotel, called Hilbert's Hotel. So many people wanted to stay there that the manager decided to build an addition. But still the hotel was always full, so he continued to add on until eventually the hotel became infinitely large. The manager was very pleased. He was particularly fond of his slogan: "We are always full—but we always have room for you."

One day a guest came to the hotel demanding a room. The receptionist pored through the books but was unable to find an empty room in the entire hotel. Luckily, the manager knew just what to do.

He gave orders for all the guests to be moved into different rooms. The guest in room 1 was moved to room 2. The guest in room 2 was moved to room 3. The guest in room 3 was moved to room 4, and so on. Everyone had a room to move to, because although there was an infinite number of guests, there was also an infinite number of rooms. Room 1 was now empty, and that room was given to the new arrival.

÷ ÷ ÷

As this story from Ray Hemmings and Dick Tahta's, *Images of Infinity* shows, the concept of infinity is one of the most counterintuitive in all of mathematics. An understanding of the basics, however, can illuminate constitutional topics to which the strictures of finite numbers never could do justice.

As a first step, we must recognize that infinity, often indicated by the symbol ∞, is not a number, not even a very, very large number. Even though sometimes it is illustrated as the end of a number line, there is no number ∞.

$$0 \quad 1 \quad 2 \quad 3 \quad 4 \quad \ldots \quad \infty$$

To grasp one of the key concepts of infinity, one must recognize that counting is really the same activity as comparing. For example, in the original U.S. flag, there is a one-to-one correspondence between the stars and stripes. Thus, it can be concluded that there are the same number of each.

The German mathematician, Georg Cantor (1845-1918), used the concept of one-to-one correspondence to compare the sizes of different infinite sets. He started with the infinite set of counting numbers, {1, 2, 3, . . .}. He then described all infinite sets that could be put into a one-to-one correspondence with the counting numbers as countable. This leads to the apparently extraordinary result that there are just as many even numbers as counting numbers. Why? Let us start with an intuitive example: There are as many even numbers as odd numbers. Even though we cannot count all the members of each set, we can see that they have the same number of members because we can put the even numbers into a one-to-one correspondence with the odd numbers:

1 3 5 7 . . .

2 4 6 8 . . .

No member of either set is left out, and we comfortably conclude that the sets of even and odd numbers have the same number of members. This technique then can be used to reach the discomfiting conclusion that there are as many even numbers as counting numbers. All we need to do is to put the even numbers into the same sort of one-to-one correspondence with the counting numbers:

1 2 3 4 . . .

2 4 6 8 . . .

For every even number, there is a counting number. No even number is unaccounted for, and, more surprisingly, no counting number is left unaccounted for either. Therefore, we can conclude that there is the same quantity of even numbers as there is of counting numbers.

This may seem problematic, because the set of even numbers is created by *removing* the odd numbers from the set of counting numbers. Therefore, shouldn't the original set have more members than what remains after the deletion? No. Infinity requires an understanding that, sometimes, the whole is not greater than the part. In fact, one definition that some mathematicians use to define an infinite set is a set "whose elements can be paired off with a subset of itself which does not contain all of the elements of the original set." To those who question this possibility, mathematical historian Morris Kline, responded, "[I]f we accept one-to-one correspondence as a basis for deciding the numerical equality of infinite collections we must agree to this seeming absurdity."

Cantor designated the countable sets, also termed denumerable sets, as having the power $\aleph_0$. Thus, the counting numbers, the odd and even numbers, even the prime numbers, all have the power of $\aleph_0$. Arithmetic involving $\aleph_0$, sometimes called the arithmetic of transfinite cardinal numbers, can look very strange.

For example, $1 + \aleph_0 = \aleph_0$. This "seeming absurdity" helps explain the story of the Hotel Infinity at the beginning of this chapter. The hotel began with an infinite number of rooms $\{R_1, R_2, R_3, R_4, \ldots\}$ filled with an infinite number of guests $\{G_1, G_2, G_3, G_4, \ldots\}$. There is a one-to-one correspondence between guests and rooms:

$$R_1 \quad R_2 \quad R_3 \quad R_4 \quad \ldots$$

$$G_1 \quad G_2 \quad G_3 \quad G_4 \quad \ldots$$

Now, let $G_0$ be the new guest looking for a room. We add $G_0$ to the set of guests $\{G_0, G_1, G_2, G_3, G_4, \ldots\}$ and find that we still have a one-to-one correspondence between the guests and rooms:

$$R_1 \quad R_2 \quad R_3 \quad R_4 \quad R_5 \quad \ldots$$

$$G_0 \quad G_1 \quad G_2 \quad G_3 \quad G_4 \quad \ldots$$

Accordingly, we can say we have proven that $1 + \aleph_0 = \aleph_0$. In the Hotel Infinity, there is always room for a new guest.

But not all infinite sets are of the same size. Cantor proved as well that the set of decimals (also known as the real numbers) was not countable. Remember, decimals can be either terminating, such as 0.123, or repeating, such as 0.142857142857 . . . (which is 1/7), or nonrepeating, such as $\pi$ (which begins 3.14159 . . . and never falls into a pattern). Cantor showed that no matter how one tries to list decimals, there always will be more that can be squeezed between any two on the list. Cantor designated the power of the set of decimals as $C$. So, even though both $C$ and $\aleph_0$ are infinite, we can conclude that $C > \aleph_0$.

To summarize, there are three relevant points about infinity that will be useful in discussing the Constitution. The first is that you never can reach infinity; it goes on forever. Second, some infinite sets are the same size, even though they do not seem as if they should be. Finally, not all infinite sets are the same size; some are larger than others.

## "From generation to generation, to the end of time, if anything human can so long endure"

> *[N]o society can make a perpetual constitution. . . . The earth always belongs to the living.*
>
> —THOMAS JEFFERSON (JULY 12, 1816)

The United States has a Constitution of infinite duration. As Charles Pinckney of South Carolina declared, "[T]his Constitution is not framed to answer temporary purposes. We hope it will last for ages—that it will be the perpetual protector of our rights and properties."

The perpetual nature of the Constitution was expressed in the Preamble, which states that the Constitution was established to "secure the Blessings of Liberty to ourselves and our Posterity." The beneficiaries of this document, "our Posterity," include not merely those living and their children, but "All the descendants . . . in a direct line to the remotest generation."

The concept of a permanent constitution was a remarkable change from the Articles of Confederation. That document, created quickly in a time of war, did not pretend to such lofty ambitions. The last article declared that "the Articles of this confederation shall be inviolably observed by every state, and the union shall be perpetual. . . ." Note that it was the "union" of the states that was perpetual, not the "Articles."

Opponents of the Constitution derided the concept of any document of perpetual durability. Noah Webster said that it was "both arrogant and impudent" for the framers to believe that they had the "perfect wisdom and probity" necessary to create a governmental structure that would last in perpetuity:

> *[T]he very attempt to make* perpetual *constitutions, is the assumption of a right to control the opinions of future generations; and to legislate for those over whom we have as little authority as we have over a nation in Asia.*

Even Thomas Jefferson, who was a supporter of the Constitution, was doubtful as to the validity of a perpetual constitution. Writing to James Madison, Jefferson wondered "[w]hether one generation of men has a right to bind another. . . ." Using actuarial tables to calculate that on average half of those of voting age, 21 years or older, would die within approximately 19 years, Jefferson concluded: "Every constitution then, and every law, naturally expires at the end of 19 years. If it be enforced longer, it is an act of force and not of right."

By 1816, Jefferson, in commenting on proposed revisions for the constitution for Virginia, revealed that he had mellowed only slightly in his antipathy to the concept of a "perpetual" constitution. He argued that, because each generation should be free to choose the form of government that it believes is most conducive to happiness, all constitutions should provide for their own periodic revision. Jefferson wrote:

> *[A] solemn opportunity of doing this every nineteen or twenty years, should be provided by the constitution; so that it may be handed on, with periodical repairs, from generation to generation, to the end of time, if anything human can so long endure.*

It would be difficult to underestimate the turmoil the United States would suffer had our Constitution mandated a "solemn opportunity" for revision every 20 years. No ruling of the Supreme Court would be authoritative for more than a brief period of time. There would have been bedlam bordering on revolution, not solemnity, had a revisionary convention been held in 1955, the year after *Brown v. Board of Education*.

≠ ≠ ≠

Nonetheless, the framers were not so arrogant as to assume that their "perpetual" document would be perfect for all time. At the Constitutional Convention, George Mason stated that "The plan now to be formed will certainly be defective, as the Confederation has been found on trial to be. Amendments therefore will be necessary, and it will be better to provide for them, in an easy, regular and Constitutional way than to trust to chance and violence."

James Iredell, who later was to serve on the Supreme Court, agreed: "The Constitution of any government that cannot be regularly amended when its defects are experienced, reduces the people to this dilemma—they must either submit to its oppressions or bring about amendments, more or less, by a civil war."

An effective amendment procedure permits a perpetual constitutional government, despite its implication that some of the provisions of that constitution are not destined to be permanent. "Happy this, the country we live in!" Iredell exclaimed. "[I]t is a most happy circumstance, that there is a remedy in the system itself for its own fallibility."

Alexander Hamilton concluded the Constitution would continue until and unless it was properly amended:

> *Until the people have, by some solemn and authoritative act, annulled or changed the established form, it is binding upon themselves collectively, as well as individually; and no presumption, or even knowledge, of their sentiments, can warrant their representatives in a departure from it, prior to such an act.*

Thus, in a somewhat Newtonian sense, the Constitution is infinite. Just as a body in motion will continue indefinitely unless some force acts on it, so will the Constitution stay in force until altered by "some solemn and authoritative act."

## Infinite Rights

The concept of infinity helps explain why some constitutional conflicts are so difficult to resolve. When dealing with finite quantities, we can decide readily which is greater. But in the realm of fundamental rights and liberties, we are dealing with freedoms of infinite value for which simple comparisons may be impossible.

In *Nebraska Press Association v. Stuart,* for example, the Supreme Court was faced with the question of whether the danger that pretrial publicity would prejudice a jury justified a ban on press coverage of a murder trial. At issue was how to resolve "the conflict between these two important rights . . . between the rights to an unbiased jury and the guarantee of freedom of the press. . . ."

The dilemma here is that the framers saw that both the right of free expression and the right to a fair jury trial were of infinite value to a free people. As was argued during the debates over ratification of the Constitution: "The Liberty of the Press . . . is a Privilege of infinite Importance . . . for which . . . we have fought and bled." Yet the right to jury trial also was seen to be of infinite importance:

> *England, from whom the Western World has largely taken its concept of individual liberty and of the dignity and worth of every man, has bequeathed to us safeguards for their preservation, the most priceless of which is that of trial by jury.*

The Supreme Court prudently avoided a battle of infinities by essentially reaffirming that each of these rights were of infinite value. Chief Justice Burger explained that, because the "authors of the Bill of Rights did not undertake to assign priorities as between First and Sixth Amendment rights, ranking one as superior to the other . . . it is not for us to rewrite the Constitution by undertaking what they declined." Mathematically, the Court decided that $\aleph_0 = \aleph_0$.

The Court was able to maintain this equality by rejecting the notion that "there is an inherent conflict that cannot be resolved without essentially abrogating one right or the other." To avoid this conflict, the Court said, trial judges must explore other methods of ensuring an impartial jury, such as change of venue, delayed proceedings, or sequestration of the jury before issuing a restraint on the press. Thus, neither right is considered greater than the other, and both can be preserved.

× × ×

Recognizing the infinite value of speech is also helpful in explaining the need to be vigilant against even the smallest restrictions on free expression. A fraction of infinity still equals infinity ($\frac{1}{2}\aleph_0 = \aleph_0$). The loss of even a fraction of the right of free expression imposes a burden of infinite scope. Thus, even a temporary ban on the publication of the Pentagon Papers by the *New York Times* and the *Washington Post* was unacceptable: "[E]very moment's continuance of the injunctions against these newspapers amounts to a flagrant, indefensible, and continuing violation of the First Amendment."

The concept of infinite rights, however, is not synonymous with absolute rights. Justice Hugo Black repeatedly argued that the language of the First Amendment, "Congress shall make no law . . . abridging freedom of speech or of the press . . . ," literally meant no law limiting any speech ever could be permitted. He contended that "The Federal Government is without any power whatever under the Constitution to put *any type* of burden on speech and expression of ideas of any kind. . . ." This absolutist position never has commanded a majority of the Court. Instead, there is general acceptance of the principle that speech can be abridged when necessary to protect some compelling governmental interest, such as preventing the publication of troop movements during war time.

Judge Learned Hand tried to capture this delicate balance in a mathematical formula. He wrote that a restraint on free speech would be constitutional only if "the gravity of the 'evil,' discounted by its improbability, justifies such invasion of free speech as necessary to avoid the danger." This can be described mathematically as $L \times P > B$, where $L$ is the loss to society if the evil occurs, $P$ is the probability of such harm resulting from the speech, and $B$ is the burden to be imposed on free speech by the government. Under this formula, a restriction will be viewed as constitutional only when $L \times P > B$, when the expected value of the harm (the product of the harm multiplied by its probability) is greater than the burden on speech.

Obviously, we should not take this approach too literally and try to calculate the burden on speech in any arithmetic way. Neither the benefits of freedom of expression nor the loss to society from its denial can be reduced to a quantity that fits into this formula.

But Hand's formula still serves a valuable function. The most significant concept inherent in this formula comes from the arithmetic truth that the product (danger × probability) is always less than the danger itself. Why? Unless an event is certain to occur, its probability is less than 1. Multiplying a number by a fraction smaller than 1 results in a product smaller than the original number. Thus, courts are reminded to gauge not only the harm arising from the sky falling down but the likelihood of such an occurrence.

This step was missed by the U.S. district court judge who imposed a prior restraint on *The Progressive* magazine's publication of the recipe for building a hydrogen bomb. The court reasoned that the government's request for an injunction should be granted because "a mistake in ruling against *The Progressive* will seriously infringe cherished First Amendment rights. A mistake in ruling against the United States could pave the way for thermonuclear annihilation for us all. In that event, our right to life is extinguished and the right to publish becomes moot."

The court's error was its failure to factor in, as Learned Hand did, whether the danger feared by the government was plausible. Mathematically, the judge failed to discount the danger of thermonuclear annihilation by the likelihood (or improbability) of thermonuclear annihilation. If we were to follow the court's example in the case of *The Progressive* and not require the government to prove the likelihood of danger, the speaker would always lose, because the transfinite harm from war and destruction must exceed the harm from a one-time loss of free speech.

There is another lesson to be gained from Hand's formula. The mathematics leads us to conclude that the greater the restraint on speech (however we choose to measure this amount), the greater the burden on the government to establish the certainty of a harm's occurring. And that leads to the proposition that any restraint on speech must be the least restrictive necessary.

In a complicated world, we must recognize that there frequently will be situations where more than one interest of infinite value is at stake. Simplistic comparisons must, therefore, of necessity give way to a far more sensitive evaluation of these competing interests.

## Abortion

Nowhere in U.S. law today is the conflict between interests of infinite value more apparent than in the issue of abortion. There is probably no way to balance all of the interests involved, and if there were such a way, I am sure that it would not involve mathematical analysis. Nonetheless, what mathematics can offer is a means to help clarify certain positions, even if only in a small way, so that each side may respect the other.

For example, my life, to me, is of infinite value. I will do whatever I can to prolong my life, to be as healthy as I can be. No amount of life insurance would cause me to drive "accidentally" off a cliff. Confronted with the demand "Your money or your life," my choice is obvious.

My children's lives, however, are worth far more to me than my own. I would not hesitate to lay down my life for my children. Again, the choice is obvious.

To translate into (cold) mathematical concepts, my assessment of the value of my own life is comparable to the number, $\aleph_0$, of counting numbers. It is infinitely greater than the value of any and all material goods. However, the value of my children's lives far transcends my own. That value is like the number, $C$, of decimals. And, as seen earlier, $C > \aleph_0$. There are two infinite values, but one is infinitely greater than the other. The imagery of infinity reflects the importance of the concerns; the power of mathematics enables us to recognize that even in the highest realms, priorities can be assigned.

Each side in the abortion debate has been guilty of occasionally trivializing the values of the other. For example, Justice Blackmun argued that a fetus prior to viability "cannot reasonably and objectively be regarded as the subject of rights or interests *distinct from,* or paramount to, those of the pregnant woman." That is not strictly true. The government's interest in "protecting potential human life" indeed can be "distinct" from its interest in protecting the rights of pregnant women. One need not decide the constitutional issue in order to acknowledge that there is something uniquely precious about fetal life.

A corresponding devaluation occurred in Justice Byron White's attack on the initial ruling in *Roe v. Wade:* "The Court apparently values *the conveniences* of the pregnant mother more than the continued existence and development of the life or potential life that she carries." Again, one need not decide the constitutional issue in order to acknowledge that there is more at stake than mere "conveniences," rather a woman's ability to make the "most basic decisions about family and parenthood." As Justice Ginsburg wrote before joining the Court, "[I]n the balance is a woman's autonomous charge of her full life's course . . . her ability to stand in relation to a man, society, and the state as an independent, self-sustaining, equal citizen."

Let us acknowledge, then, that it is not only noncontradictory but plausible to accept the view that neither the fetus's nor the woman's interests are trivial. Indeed, they are both of infinite importance. That hardly resolves the issue. Recalling that some infinities are greater than others ($C > \aleph_0$), the ultimate question will be whether the greater value is to be assigned to the fetus's life or to the woman's ability to control her body and destiny. That, of course, is not a mathematical question but the heart of the abortion debate.

So what has been gained in using transfinite mathematics to analyze the abortion question if we still are left with the truly difficult legal, political, and ethical questions? Perhaps just a framework for understanding that those whose conclusions differ fundamentally from our own need not be viewed as also denying the very essence of our views of life, personhood, and God. That understanding might permit, in turn, a more civil and a more sympathetic dialogue despite the broad areas of disagreement. And maybe that is not such a small thing, after all.

# 9

# The Incomplete Constitution

> **Our world is endlessly more complicated than any finite program or any finite set of rules. You're free, and you're really alive, and there's no telling what you'll think of next, nor is there any reason you shouldn't kick over the traces and start a new life at any time.**
>
> **—*Rudy Rucker,* Mind Tools *(1987)***

## Russell's Barber

Bertrand Russell told the story of a barber who lived in a small village and was very proud of the fact that he was the only barber in town. To emphasize that distinction, he wanted to post a sign that read "I will shave everyone in town who shaves." It occurred to him, though, that there were probably some in the town who shaved themselves. After some thought, he put the following sign on the door of his shop: "I will shave anyone who does not shave himself, but I won't shave anyone who does."

The next morning, the barber awakened in a state of distress. He was about to shave himself, when he remembered his sign. He realized that if he shaved himself, he would violate his pledge to shave only those who did not shave themselves. But if he did not shave himself, he also would violate his pledge, this time the first part of his guarantee that he would shave all those who did not shave themselves.

The confused barber stared into his mirror for a long long time.

## Gödel's Incompleteness Theorem

In 1928, one of the leading mathematicians of the twentieth century, David Hilbert (whose eponymous hotel began the chapter on infinity), issued a challenge to the mathematical community. He wanted someone to create a consistent and complete formal logical system, one that would determine the truthfulness of every mathematical statement. Consistency required that the system never prove that both a statement and its negation were true. Completeness meant that a finite set of axioms and rules would be able to prove or disprove any mathematical hypothesis.

In 1930, a 24-year-old mathematician, Kurt Gödel, announced an unexpected end to the challenge. Gödel had proven that no such system could ever be created, that within any system of the kind envisioned by Hilbert, there always would exist true mathematical statements that were unprovable. Such systems were fated to be incomplete.

A formal system begins by defining a few symbols, which are considered the vocabulary for that particular system. These symbols can be combined into strings of symbols, and the few axioms of the system are created using these initial strings. The system also contains specific transformation rules, which tell how strings may be turned into new strings. A proof in such a formal system begins with the axioms and, through repeated transformation, ends with a theorem, some desired new string of symbols.

Gödel's first step was to take the most basic symbols of mathematics and logic (such as 0; *f*, the symbol for "the successor of"; and ~, the symbol for "not"; among others) and represent each with its own unique number (usually called its Gödel number). The symbols can be combined into strings of mathematical statements, with each string then given its own unique Gödel number. Using Gödel's rules of transformation, strings can be transformed into new strings. A collection of strings is called a proof schema, because the last string in the collection is considered proven if it results from the correct transformations of the preceding strings in the collection.

The next major innovation by Gödel was to figure out how to encode metamathematical statements in the same systems as mathematical ones. A metamathematical statement is one that is about the system, rather than in the system. To say "$7 \times 3 = 21$" is a mathematical statement within the system; to say "There is no even prime number other than 2" is a metamathematical statement, since it talks *about* the number system.

This same distinction occurs in everyday language. The statement "The sky is blue" uses words to make a statement. By contrast, the statement "A sentence needs a subject and a predicate" is a metastatement referring to the system of language we are using. A self-referential metastatement talks about itself, as in "This sentence contains a subject and a predicate."

Gödel's breakthrough was to create an encoded metamathematical statement that was also self-referential. Gödel found a way to encode "This statement is not provable within this system."

As described by Gödel, "We now come to the goal of our discussions." He had taken for his inspiration the classic Liar's Paradox. Imagine being confronted with the task of judging the truthfulness of the self-referential declaration "This sentence is false." We cannot treat the sentence as true, since the sentence *said* it was false. If, instead, we try to treat the sentence as false, we also run into problems, since *then* the sentence, which declared itself false, would turn out to be true. Thus, we cannot treat the sentence as either true or false without a fatal contradiction. A similar contradiction immobilized the barber in the parable by Bertrand Russell which began this chapter.

Gödel constructed a metamathematical statement within his formal system that declared, in effect, "This proposition is not provable within this formal system." What gives the statement the power to transcend the problem of paradox is the fact that, unlike the Liar's Paradox, it does not refer to falsity but to a lack of provability.

By separating the two, Gödel was able to complete his proof. If his proposition were in fact provable, it would be false. This would be an unacceptable contradiction because, by definition, the system can prove only truthful statements. On the other hand, if the statement were true, it would be merely unprovable, which would create no contradiction. Thus, we can conclude that the statement declaring its own unprovability is true. And that means that we have found at least one statement that is both true and not provable within the system. Gödel was able to show further that no matter how many axioms one adds to a system, it is always possible to find other statements that are true but unprovable within the system. And thus, the formal system is destined to remain incomplete.

It must be pointed out, however, that true statements that are not provable within one particular system may well be provable in another system. In fact, the truthfulness of Gödel's metamathematical statement was proven by logic outside his formal system. We know that any second system also will be incomplete but will have different statements beyond its capacity. Thus, there is no escaping the fundamental dilemma of incompleteness.

= = =

To travel from the sparse definitions of a formal logic system to the wide-ranging world of law requires an enormous leap.

Douglas Hofstadter, in his breathtaking book, *Gödel, Escher, Bach: An Eternal Golden Braid,* issued a warning to those who would try to apply Gödel's theorem directly to other disciplines: "It would be a large mistake to think that what has been worked out with the utmost delicacy in mathematical logic should hold without modification in a completely different area."

It is impossible to apply Gödel's analysis directly to law to prove that, in any sense, law is incomplete. Law professor Mike Townsend points out, "[O]ne would have to provide a legal language, which in turn requires specifying an alphabet and the set of legal formulas; a set of legal axioms; and a set of rules of legal inference. . . . The resulting legal formal system must [include] something like [Gödel's] encoding properties. . . ." The necessary prerequisites for Gödel's theorem just do not exist in law.

It also would be a mistake to treat Gödel's theorem as representing proof that mathematical, logical, or legal systems are indeterminate and uncertain. The fact that not all statements are provable within a certain system establishes neither that nothing is provable nor that a given statement is not provable in any system. Even after Gödel, mathematicians prove theorems.

≥ ≥ ≥

Yet, those studying the Constitution still can gain useful insight by considering various aspects of Gödel's proof and theorem. To begin with, just as Gödel had to distinguish between mathematical and metamathematical statements, we must be aware of the difference between constitutional rules and those meta-constitutional rules which describe how the rules are to be derived. A constitutional rule, for example, would describe the right of an individual to free expression or would prohibit [or permit] certain forms of discrimination. A meta-constitutional rule would describe how the Court decides a case.

Many of the more important metarules reflect the passive virtues of insuring that political experimentation will not be foreclosed unnecessarily by a constitutional decision of the Supreme Court. Thus, the Court "will not pass upon the constitutionality of an act of Congress . . . unless such adjudication is unavoidable. . . ." When the Supreme Court interprets a statute narrowly so as to avoid reaching a constitutional issue, the Court is implementing a metarule.

There can exist multiple levels of metarules. These meta-meta-constitutional rules tell the Court how to apply its rules regarding how to decide cases. For example, the Court has stated that its metarule on avoiding constitutional decisions would not be applied in an equal protection case, when the result would be relitigation of "an insubstantial state issue." The result of this meta-meta-constitutional rule was that the Court was able to issue a ruling on the constitutional merits of the case.

≠ ≠ ≠

Other than for classification purposes, the distinction between constitutional rules and higher-level meta-constitutional rules is important for recognizing one of the fundamental distinctions in constitutional jurisprudence: the difference between the question of *who* gets to decide whether a given action is consistent with constitutional norms and the question of *whether* that action is consistent with those norms. The "who" question is properly viewed as the meta-constitutional question which must be answered before the constitutional "whether" question can be resolved.

*Marbury v. Madison* was the first case in which the Supreme Court struck down a law of Congress as unconstitutional. In its most widely quoted passage, Chief Justice Marshall declared, "It is emphatically the province and duty of the judicial department to say what the law is." The Court then said that a statute authorizing the Court to issue a writ of mandamus (to direct President Jefferson to hand over the commission to Marbury) was unconstitutional because it expanded the Court's jurisdiction beyond that established in the Constitution.

Just 15 years later, the Court in *McCulloch v. Maryland* upheld Congress's creation of a national bank as a "necessary and proper" exercise of legislative power but refused to determine whether the bank was "necessary." The Court stated that as long as a measure was "calculated to effect any of the objects entrusted to the government," then the "degree of its necessity . . . is to be discussed in another place."

*Marbury* and *McCulloch* reflect two different resolutions of the metalevel constitutional inquiry. In *Marbury* the metaquestion was who decides whether the authorization of the writ of mandamus is constitutional: Congress or the Court? Having decided that the answer was the Court, Justice Marshall then went on to answer the constitutional question in the negative. In *McCulloch,* the Court decided the "who" question differently, stating that, in most circumstances, Congress has the discretion to determine what is "necessary and proper" under the Commerce Clause. Congress, then, had the authority to decide that a bank was necessary.

+ + +

When the Supreme Court strikes down a law, it will have made determinations on two distinct levels. On the first level is the meta-constitutional ruling that the merits of the particular constitutional question should be decided by the judicial branch. On the second level is the constitutional ruling that the law does indeed conflict with the Constitution.

The distinction between the two different levels of judicial determination has not always been recognized by the Court itself. For example, in *U.S. v. Nixon* the Court held that presidents have only a qualified privilege to keep information secret and that it is the province of the judiciary to decide whether other interests outweigh the need for presidential confidentiality. In rebutting President Nixon's claim that the Constitution provides the president with an absolute privilege, Chief Justice Burger wrote: "Many decisions of this Court, however, have unequivocally reaffirmed the holding of *Marbury* that 'It is emphatically the province and duty of the judicial department to say what the law is.'"

However, *Marbury's* holding does not provide the answer to the claim of absolute presidential privilege, because the law involved is the meta-constitutional question of who should decide what presidential information is too important to disclose. In other words, there were two distinct questions in the *Nixon* case. The first question (the metaissue) asks, which branch decides whether it is necessary to keep specific presidential information secret? The second question asks, should this specific presidential information be kept secret? The power of judicial review explained in *Marbury,* therefore, need not foreclose a discussion of the issue of which branch the Court should deem responsible for making this particular determination.

There are at least two levels to every constitutional challenge. Before the judicial department can rule on the substance of an issue, it first must decide which branch is responsible for making that substantive determination.

This issue was at the heart of the Supreme Court's decision in *Bush v. Gore*. The Florida Supreme Court had ordered hand recounting of all ballots on which the voting machines had failed to detect a vote for president but had not provided a specific standard for determining the intent of the voters. The U.S. Supreme

Court held that such a recount would violate the Equal Protection Clause because different standards would be used by different counties, and thus the Court ended the recount and settled the 2000 presidential election once and for all. Before deciding the equal protection issue, however, a majority of justices needed to resolve the metaquestion, whether the Supreme Court should have the task of determining the fairness of the selection of a state's electors: "When contending parties invoke the process of the courts, however, it becomes our unsought responsibility to resolve the federal and constitutional issues the judicial system has been forced to confront."

An angry four-justice dissent declared that the Supreme Court had answered the metaquestion incorrectly and never should have heard the case. In the words of Justice David Souter: "If this Court had allowed the State to follow the course indicated by the opinions of its own Supreme Court, it is entirely possible that there would ultimately have been no issue requiring our review, and political tension could have worked itself out in the Congress."

An analogous metaissue can be seen in the question of whether a congressional declaration of war was necessary prior to Operation Desert Storm. While the president is commander in chief, the Constitution gives Congress the power to declare war. But who has the (meta-constitutional) job of deciding what kind of conflict qualifies as a war? The Court must first decide to whom the law of the Constitution assigns this definitional task: Is this a political question to be decided by the president or Congress, or is it a judicial question to be resolved by the courts? The Supreme Court has yet to answer this meta-constitutional question. If it ever does attempt to resolve this issue, the Court only would confront the question of whether a particular conflict was a war if the answer to the metaquestion was that the Court should decide when a conflict becomes a war.

## "This Title is Self-Referential."

Self-reference may arise in legal analysis, just as it can in a formal logic system. Sometimes the existence of self-reference creates no great difficulty, while at other times it may lead to problems similar to those deliberately created by Gödel.

Take the issue of whether a constitutional decision should apply to cases that arose prior to the announcement of the decision. On June 13, 1966, in *Miranda v. Arizona,* the Supreme Court announced that it would throw out convictions that were obtained with evidence garnered through improper police questioning of a suspect. Shortly thereafter, the Court held that this rule of constitutional law would be "available only to persons whose trials had not begun as of June 13, 1966." What about Miranda himself? Since his trial surely had begun before the Supreme Court ruled on his case, this would mean that the beneficial ruling was not available to him. But how could his conviction not be overturned after he won his case? At first, the Court's way out of this dilemma was to treat the person who brought a successful constitutional challenge as a one-person exception to the rule on non-retroactivity. Later, the Court decided that the exception was unfair and broadened the rule to permit retroactive application to all those with cases still subject to direct appellate review when a new constitutional rule was announced.

≠ ≠ ≠

The use of precedent, *stare decisis,* to decide new cases also may create questions of self-reference. As Justice Scalia has argued, "*stare decisis* ought to be applied even to the doctrine of *stare decisis*. . . ."

As an initial matter, the source of the doctrine of *stare decisis* must come from outside the system of earlier case law. It is certainly true that "The doctrine of precedent cannot be authoritatively supported by references to precedent; it cannot pull itself up by its own bootstraps." Thus, if a court wishes to utilize precedent from older cases to support newer decisions, such practice needs to be justified from outside the system, either by legislative mandates, a court rule of procedure, or a normative determination that precedent is an appropriate way for a court to proceed.

But dealing with precedent about precedent creates all sorts of problems. In *44 Liquourmart, Inc. v. Rhode Island,* a four-justice plurality implied that the Court should give less deference to decisions reached by a 5 to 4 vote than those reached by wider majorities. Imagine what would have happened had that plurality obtained one more vote. We would have had a 5 to 4 Supreme Court decision, stating a rule of decreased deference for 5 to 4 majority decisions (call it case 1). Now, suppose that the next year, case 2 comes up. Say that one of the justices in that majority has had a change of heart and now believes that all majority opinions should have equal precedential value. What should be the precedential value of case 1 to this justice? Would it be appropriate to give decreased deference to the doctrine of decreased deference? But if you believe in equal deference, how can you not give equal deference to all former decisions, including case 1?

× × ×

A similar dilemma arises with disputes between different branches of government over constitutional interpretation, as illustrated by the following near miss from U.S. history. In 1867, in response to fears that the Supreme Court would strike down Reconstruction laws, the House of Representatives voted to require a two-thirds vote of the justices in order for any federal statute to be declared unconstitutional. This bill died in the Senate. But what if it had become law? Suppose the law was challenged, and the Supreme Court voted by a 5 to 4 margin to declare it unconstitutional. How would we know the status of the law? The Court would be saying that the law was unconstitutional, but the law would be saying the opposite—that it had not been declared unconstitutional by the Court. We could go around in circles, except that, under our current system, the Court has the last word, because "the federal judiciary is supreme in the exposition of the law of the Constitution, and that principle has ever since been respected by this Court and the Country as a permanent and indispensable feature of our constitutional system." As one commentator put it, "*Marbury* put a firm halt to the infinite regress."

However, this principle of judicial supremacy ultimately must be derived from outside the system, from outside the Constitution itself. The issue of who is the ultimate arbiter is, in essence, "logically antecedent to the written constitution. Even a written constitution explicitly specifying its authoritative interpreter would rest on a constitutional understanding about who should be the authoritative interpreter of *that* provision." Insofar as the U.S. Constitution lacks an explicit provision stating who is its authoritative interpreter, "[I]t is even clearer that such a specification is a background decision not dependent on what is specified in the very document whose method of interpretation is at issue."

= = =

Another self-referential issue involves the amending of constitutional provisions that limit their own amendment. The Constitution prohibits amendments that would alter the requirement that states have equal voting in the Senate; the original Constitution also barred amendments that would have permitted Congress to limit the importation of slaves before 1808. Just before the Civil War, Congress passed and sent to the states for ratification a proposed constitutional amendment, known as the Corwin Amendment, after its sponsor Representative Thomas Corwin of Ohio. The Corwin Amendment would have prohibited any future constitutional amendment from authorizing Congress to outlaw slavery.

The amendment only garnered two state ratifications, but would it have been meaningful even had it received the three-quarters required? How can an amendment keep itself from being amended? Would it have been possible, for example, for the drafters of the eighteenth Amendment (prohibition) to have prevented its own eventual repeal by the twenty-first Amendment?

One logical solution to the self-referential problem of constitutional amendments limiting their own future amending is to create a legal theory of types. Mathematically, the theory of types classifies sets by the types of members they contain. For example, a type 1 set might consist of only individual objects (the set of whole numbers or the set of even numbers, for example). A type 2 set would be able to include sets of sets. Thus a type 2 set could include the type 1 sets as members. A type 2 set might be the set of all sets containing even numbers. The heart of the theory of types is the requirement that a set only can be a member of a set of a higher type. Thus, many self-referential mathematical paradoxes are eliminated because a set cannot contain itself.

The legal analogue to the theory of types would create a similar hierarchy: At the bottom rung might be agency regulations, higher would be statutes, and finally at the top, would be the Constitution itself. Under the legal theory of types, a ban on altering a given provision would be deemed effective only if that restriction were contained in a provision of a higher constitutional type. Thus, a statute can bar the change of a regulation, and a constitutional provision can prevent the changing of a statute. But a statute cannot make itself un-repealable, and a constitutional amendment, such as the Corwin Amendment, cannot bar its own future repealing. The ban on altering the equal suffrage of states in the Senate might well be saved, however, if we consider the original Constitution as being of an even higher type than subsequent amendments. Thus, subsequent amendments can and do change either the original document or other amendments, but only if such changes were not prohibited in the original document.

## Constitutional Incompleteness

The Supreme Court has been able to avoid many of the pitfalls created by constitutional self-reference through its accepted authority to decide "what the law is." Occasionally, though, the Court has failed to see that the underlying issue it faces presents just such a pitfall.

In 1988, in *Morrison v. Olsen,* the Supreme Court upheld the creation of an independent counsel to investigate and prosecute alleged wrongdoing in the executive branch. In August 1994, under a slightly revised independent counsel law, Kenneth Starr was appointed by a panel of three federal judges to serve as the chief prosecutor investigating President Clinton. He was not appointed by the attorney general or by any other executive branch official. Furthermore, unlike other prosecutors, Starr could not be fired at the sole discretion of the president but only if it could be proven to a court that he was performing his prosecutorial job badly. Despite such restrictions on the normal operations of the executive branch, the Court ruled in *Morrison* that the independent counsel law did not violate the separation of powers.

Justice Scalia, as the sole dissenting voice, argued that the law should be viewed as unconstitutional simply because criminal prosecution is a purely executive function and the statute deprived the president of complete control of that function. Moreover, Justice Scalia's dissent pointed out that the practical effect of the law was to weaken the president and eliminate the normal political check on prosecutorial abuse.

The majority opinion never addressed these points. Instead, the Court simply announced that the law was constitutional because the traditional complete control of the prosecutor by the president was not "so central to the functioning of the Executive Branch. . . ." This is a dangerous rationale, implying that there is no identifiable limit on the ability of Congress to remove any sort of executive function from the president's control. It creates the risk that more and more power would go to the legislature and the presidency would be severely weakened.

The Court's opinion would have been much stronger, and narrower, had it been focused on the only legitimate rationale for limiting the executive control of prosecution: How can the executive branch, in charge of investigating misdeeds, investigate itself? Almost a decade before *Morrison,* Douglas Hofstadter had presciently foreseen this sort of situation as a "strange loop," analogous to the paradoxical tangle created in Gödel's theorem.

The independent counsel law creates a second self-referential strange loop. As revealed by the excesses of the Starr investigation, if we cannot trust the chief executive to investigate himself, how can we expect the independent counsel to police himself? The problem is that, at some point, we will be forced to confront the self-referential dilemma of some persons being responsible for policing themselves.

A judicial opinion, backed by such insight, might begin this way:

*It should not be surprising to find that the Constitution, a system of rules, has trouble resolving certain issues of self-reference. Our constitutional system allocates responsibility by deliberately combining the contradictory concepts of separation of powers and checks and balances. At some point, these ideals clash. The issue of executive branch self-investigation is exactly the sort of self-reference that might be expected to strain the logic of this governmental structure. If we honor the separation of powers, there is no effective check on the inherent conflict of interest from the executive branch's investigating itself. If, however, we accept the check of an independent counsel, the executive function of prosecution is no longer separate from the other branches. The Constitution and Federalist Papers are silent on this matter. Perhaps this issue points to a spot where the Constitution is incomplete and where application of traditional rules leads to paradox or contradiction.*

From this point, the ultimate ruling is far from inevitable. The independent counsel law could be upheld on the theory that the Court will respect the choice made by Congress to address a serious danger of otherwise unchecked abuse when the Constitution does not resolve the problem expressly. Alternatively, the Court could strike down the law, saying that, as self-referential problems never can be avoided completely, we should settle for Justice Scalia's conclusion that, while "the separation of powers may prevent us from righting every wrong, it does so in order to ensure that we do not lose liberty." Resolving this difficult conflict will be made easier if we recognize that we should not expect our, or indeed any, constitutional structure to provide all the answers.

≥ ≥ ≥

Not every instance of self-reference will so tax our system. From *Marbury* on, the Supreme Court has ruled on the scope of its own jurisdiction. But even in this relatively stable area, complicated issues may surface.

In *City of Boerne v. Flores,* the Court turned back an attempt by Congress to use the fourteenth Amendment to reverse the effect of a recent Supreme Court ruling. At issue was the Religious Freedom Restoration Act of 1993, which prohibited both the federal and state governments from imposing substantial burdens on religious practices unless the government could demonstrate a compelling interest and prove that no less restrictive means could accomplish its purpose. This law was a direct response to a 1990 Supreme Court decision, *Employment Division, Department of Human Resources v. Smith*. In *Smith,* the Court upheld Oregon's law prohibiting the use of peyote, even though the ban prevented the ritual use of the drug as part of the religious practices of members of the Native American Church. The Court ruled that the Free Exercise Clause of the First Amendment permitted a generally applicable law to burden religious activity, even if the government did not offer a particularly important justification.

The Court in *Flores* struck down the Religious Freedom Restoration Act as an unconstitutional congressional attempt to "decree the substance" of a constitutional provision. In doing so, the Court was far more attentive to separation of powers concerns than it had been in *Morrison*. The Court admonished that "Our national experience teaches that the Constitution is preserved best when each part of the government respects both the Constitution and the proper actions and determinations of the other branches." Without a note of irony, the Court warned of the threat to our system "[i]f Congress could define its own powers. . . ." The Court seemed not to notice or be concerned that in deciding this case it was the Court itself which was acting to "define its own powers."

÷ ÷ ÷

The Court has recognized on other occasions that there is a constitutional check on judicial power which is beyond judicial control: impeachment. The question was whether a federal judge facing impeachment could challenge the constitutionality of the procedures being used by the Senate to remove him from office. The Supreme Court refused to address the merits of the Senate's procedures, and thereby avoided the entanglement of another strange loop: "[J]udicial involvement in impeachment proceedings, even if only for purposes of judicial review, is counterintuitive because it would . . . place final reviewing authority with respect to impeachment in the hands of the same body that the impeachment process is meant to regulate. . . ."

Other constitutional strange loops may arise when Congress rules on itself. For instance, the Constitution authorizes the House of Representatives and the Senate each to "be Judge of the Elections, Returns and Qualifications of its own Members." When the House tried to bar Representative Adam Clayton Powell on grounds of moral character, the Court held that neither legislative body was permitted to add its own qualifications to the three—age, citizenship, and residency—contained in the Constitution. This ruling avoided the dangers inherent in self-regulation, which James Madison had referred to as "vesting an improper & dangerous power in the Legislature." As Hugh Williamson of North Carolina warned during the Constitutional Convention, "Should a majority of the Legislature be composed of any particular description of men, of lawyers for example, . . . the future elections might be secured to their own body."

A more difficult problem arises when Congress attempts to exempt itself from its own laws. One of the principles behind the separation of powers is that by detaching the legislative from the executive power, those who pass the laws cannot avoid the force of the law. In theory, Congress will be reluctant to pass a tyrannical law if it knows it also will be subject to the law's requirements. As early as 1689, John Locke had written that it would be "too great a temptation to human frailty" to permit legislators to "exempt themselves from the Obedience to the Laws they make, and suit the Law . . . to their own private advantage, and thereby come to have a distinct interest from the rest of the community, contrary to the end of Society and Government."

Nonetheless, for years Congress exempted itself from a wide range of employment legislation, including the 1964 Civil Rights Act, the Family and Medical Leave Act, and the Occupational Safety and Health Act. Finally, in 1995, Congress passed the Congressional Accountability Act, which subjected Congress to these and other laws of general applicability. It is not clear whether the Constitution required that Congress treat itself no differently from other employers. But in this case the loop of Congress regulating itself *with* others provides a protection against arbitrary lawmaking by forcing those who write the laws to live with their consequences.

≠ ≠ ≠

While requiring that Congress burden itself along with others avoids one kind of danger, barring Congress from conferring a special advantage on itself prevents another. Specifically, the most recent amendment to the Constitution, the Twenty-Seventh Amendment, prevents members of Congress from raising their own salaries until after an intervening election. This amendment presented a simple technique to avoid a recursive benefit ("I vote myself a raise"), because there is no guarantee that a member will survive reelection and receive the benefit.

In fact, the ratification of the Twenty-Seventh Amendment illustrates a far more serious problem caused by self-reference. Specifically, who will determine the validity of a constitutional amendment? The Twenty-Seventh Amendment was first sent to the states in 1790, along with the original Bill of Rights. It took

202 years for two-thirds of the States to ratify it. Is that too long? Before we know *if* a proposed amendment would lapse if it is not passed in a reasonable time, we first must decide which branch of the Federal Government should decide such an issue.

Officially, the formal determination of the validity of the amendment came from an executive branch employee, the national archivist. According to current law, an amendment becomes valid on the archivist's certification that the appropriate number of states have ratified it. Arguably, of the three branches of government, the executive is the *least* appropriate one for this task. First of all, Article V, which provides for amendments, deliberately provided no role for the president. Secondly, determining whether the appropriate procedures were followed to make a law binding is not an executive function.

Since determining the validity of laws is a judicial function, it might seem most appropriate that the Supreme Court make the final determination of the validity of the enactment of a constitutional amendment. However, the last time the Supreme Court was asked to rule on such validity, in the 1939 case of *Coleman v. Miller,* a fractured Court ruled that this determination was a nonjusticiable political question. The case arose out of the attempt to ratify the Child Labor Amendment, which would have undone the effect of various Supreme Court cases and permitted Congress to regulate child labor. Although it probably is correct to view *Coleman* as lacking a majority opinion, the case has come to be interpreted as standing for the proposition that Congress has the ultimate authority to determine the legitimacy of the ratification of amendments. This rationale subsequently was explained by Justice Powell:

> *The proposed constitutional amendment at issue in* Coleman *would have overruled decisions of this Court. Thus, judicial*

> *review of the legitimacy of a State's ratification would have compelled this Court to oversee the very constitutional process used to reverse Supreme Court decisions. In such circumstances it may be entirely appropriate for the Judicial Branch of Government to step aside.*

Thus, the argument goes, to avoid the strange loop of a judicial decision on the validity of a process designed to reverse a judicial decision, the Court prudently will avoid involvement. But this argument proves too much. Amendments also have been designed to force or restrain executive or congressional action. Indeed, the twenty-seventh Amendment is designed to act solely on Congress. Accordingly, judicial review of the amending process is only a strange loop if judicial conduct is being restrained by a particular amendment.

But what if Congress proposed an amendment to increase its own power? Or what if one Congress wisely proposed an amendment to restrain a dangerous congressional activity, but a subsequent Congress opposed the limitation? In either case, if Congress is the target of a constitutional amendment, it would be a strange loop for Congress to make the final determination concerning the process designed to control congressional determinations. Indeed, many at the Constitutional Convention voiced the concern that "It would be improper to require the consent of the Natl. Legislature, because they may abuse their power, and refuse their consent on that very basis."

Thus, we apparently have ruled out all three branches of the federal government as the appropriate final arbiter of the validity of the ratification of constitutional amendments. Perhaps we have just located an inevitable incompleteness in the constitutional structure. The amendment ratification decision is self-referential. It involves interpreting the meaning of the constitutional provision relating to changes in the meaning of the Constitution. Moreover, whichever branch is responsible for determining the validity of the amendment process may end up ruling on an amendment that was designed to curb its own power. This instance of double self-reference is, at the very least, a likely candidate for pushing a rule-based system to its logical limit.

But this does not mean that the problem is unresolvable. Undoubtably, someone will act to decide the validity of each amendment, even if the resolution is logically imperfect. The incompleteness created by this self-referential situation might even be seen as a blessing, creating an opportunity for creative problem solving. As one optimist noted, "It may well be the path of practical wisdom to preserve the logical gap as one of those features which allow a saving flexibility to the structure of society in its legally ordered aspect. . . ."

≠ ≠ ≠

Kurt Gödel was born on April 28, 1906, in what is now the town of Brno in the Czech Republic. He came to the United States for good in 1940 and applied for citizenship eight years later. To prepare for his oral citizenship exam, he studied the Constitution. During his review, he realized that while the Constitution provided for a democratic government, its amendment procedures permitted the theoretical elimination of its structural and substantive protections of liberty.

On April 2, 1948, Gödel went to the immigration office, accompanied by his good friends Albert Einstein and Oskar Morgenstern (a mathematician of equal stature). The interview proceeded smoothly, until the immigration official made the mistake of declaring that, while Germany had suffered under an evil dictatorship, such a fate was not possible in the United States.

# 10

# Constitutional Chaos

**A very small cause which escapes our notice determines a considerable effect that we cannot fail to see, and then we say the effect is due to chance. . . . [I]t may happen that small differences in the initial conditions produce very great ones in the final phenomena. A small error in the former will produce an enormous error in the latter. Prediction becomes impossible, and we have the fortuitous phenomenon.**

***—Henri Poincaré (1908)***

It is very hard not to think in a linear fashion, either mathematically or otherwise. Even without using the name, most people tend to have an intuitive sense of the relationships that are inherent in linear functions. Unfortunately, that intuition leaves a person disinclined to deal with the complex patterns and less-than-predictable results of a multifaceted, nonlinear world.

Relationships in mathematics often can be described as functions. A function is a mathematical machine; you put in the input and it produces an output. If we have a doubling function, when you put in 2, you get 4; put in 3, you get 6; and so on. The mathematical notation of this particular relationship is $f(x) = 2x$: put in $x$ and it produces twice that $x$.

One of the easiest ways to understand how a function treats a wide array of inputs is to draw a graph of the functional relationship. Below are two graphs, one for $f(x) = 2x$, the other for a slightly more complicated function, $f(x) = -x + 9$.

The most noticeable aspect of these graphs is that the resulting picture in each is a straight line. That is why functions of this sort are called linear functions, and equations that express such relationships are called linear equations. In the first graph, the outputs increase along with the inputs. In the second, the outputs have an inverse relationship to inputs, declining as $x$ rises. Up or down, in both graphs the relationships are linear.

One significant aspect of linear functions is how well they represent the concept of proportional change. In each equation,

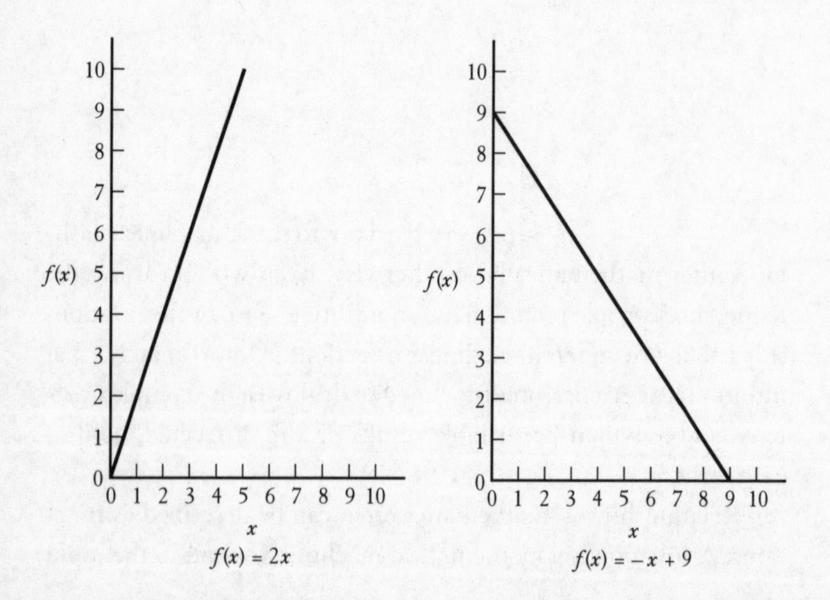

the bigger the change in input, the bigger the change in output. A small increase in input $x$ leads to a small increase in the first output and a small decrease in the second output. It is easy to see that a large increase in input will result in a correspondingly large change in output.

This translates well to much of our day-to-day lives. A slight turn of a knob on a stereo results in a slight increase in the volume of the music. The more television you watch, the less work you complete.

In reality, however, we do not live in a linear world. Countervailing forces can work either to diminish or increase our outputs. Doubling the capacity of electric wires does not double the amount of electricity they carry, due to resistance and other factors. Economies of scale and the law of diminishing returns guarantee that not all changes in production or supply will result in a proportionate change in cost or demand.

Up until the 1960s, the general feeling was that many nonlinear processes could be treated as linear over a small enough time interval. It also was presumed that the nonlinear problems that would present the most difficulty were those of greatest complexity.

The introduction of chaos theory into the mathematical and scientific community forever altered these expectations. *Chaos* has a precise mathematical meaning that is different from its common English usage. To say that either politics in the Middle East or my office desk is in chaos is not a mathematical observation.

Mathematically, chaotic systems are dynamic, meaning that they change over time. One major characteristic of chaotic systems is the concept of feedback, or iteration. This means that your next step is determined by your current status. A simple, nonchaotic example of iteration would be if our doubling function used its most recent output as its next input: Putting in 2 gives 4, putting that 4 back in gives 8, putting that 8 back in gives 16, and so on.

The next important characteristic of a chaotic system is that it displays "sensitive dependence on initial conditions." This means that the slightest difference in your starting point results in a radically different journey: "[T]he behavior of systems with different initial conditions, no matter how similar, diverges exponentially as time goes on."

A third significant characteristic trait of a chaotic system is that, again over time, future steps will take you in every direction. Outputs will rise and fall, by small and great amounts. A system with this characteristic is known as topologically transitive.

A simple example of a chaotic system can be seen in one of the most elementary biological models, relating the population of a species from one time period to the next. Populations generally do not grow indefinitely. If there is a high population density, the numbers tend to decrease, due to realities such as overcrowding and lack of food. Conversely, if there is a low population density, the population tends to grow. One generalized formula for these boom-and-bust cycles is $X_{t+1} = aX_t(1 - X_t)$, where $X_t$ is the population density at the preceding time period and $X_{t+1}$ is the population density at the next time period. The coefficient *a* represents the steepness of the growth curve, the growth parameter.

Regardless of the initial density, at low levels of *a* ($1 < a < 3$) the population will vary for a while, then settle down to a fixed number and stay there. For slightly higher values of *a*, after a period of variance, the population density oscillates between two fixed values. But beyond a certain value of *a* ($a > 3.57$), the period of variance never ends. There is a permanent lack of identifiable pattern.

Over time, the graph of population density for the three different values of *a* will resemble the following:

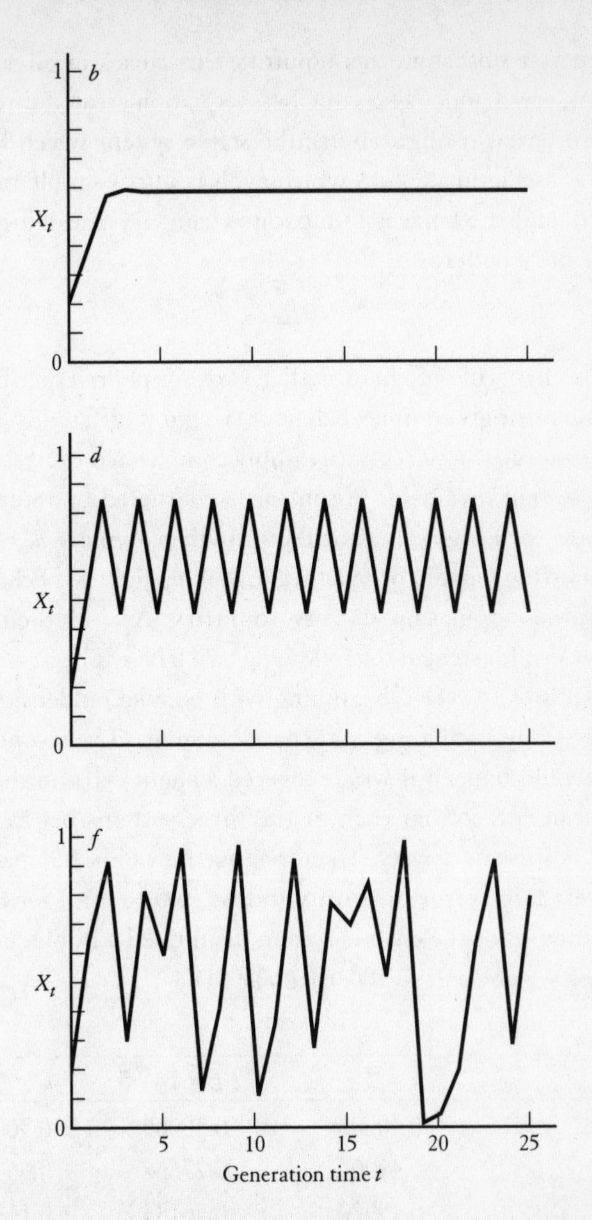


*Figure from Robert May, "When two and two do not make four: non-linear phenomena in ecology,"* Proceedings of the Royal Society of London *(1986), B228, 241, 246.*

These results show that not all systems, not even all nonlinear systems, are chaotic. The same basic equation created three strikingly different configurations: the stable system which becomes flat, the oscillating system which evolves into a simple repetitive pattern, and the hopelessly intricate chaotic system which never settles into a pattern.

= = =

The first rule of chaos is that very simple relationships can yield surprisingly complex behavior. To put it differently, complicated results do not necessarily imply complicated causes.

A second invaluable lesson can be garnered by returning to the three differing values of the growth parameter *a,* only this time showing what happens if we make the slightest of changes in our initial population density (our first $X_t$). Mathematician Franco Vivaldi created the following two tables. The first displays the results of 15 cycles, beginning with population density of 0.4, for the steady, oscillating, and chaotic systems. The second shows what would happen if we discovered a slight error in the initial measurement, so that each of the three systems began from a slightly different density. To emphasize his point, he imagined a relatively large error in density (off by 0.05 to 0.35) for the first two systems; for the chaotic system, he posited a minuscule error in density—one part in 100,000 (0.00001).

| $t$ | $a = 2$ | $a = 1+\sqrt{5}$ | $a = 4$ |
|---|---|---|---|
| 0 | 0.40000 | 0.40000 | 0.40000 |
| 1 | 0.48000 | 0.77666 | 0.96000 |
| 2 | 0.49920 | 0.56133 | 0.14360 |
| 3 | 0.50000 | 0.79684 | 0.52003 |
| 4 | 0.50000 | 0.52387 | 0.99840 |
| 5 | 0.50000 | 0.80717 | 0.00641 |

| | | | |
|---|---|---|---|
| 6 | 0.50000 | 0.50368 | 0.02547 |
| 7 | 0.50000 | 0.80897 | 0.09928 |
| 8 | 0.50000 | 0.50009 | 0.35768 |
| 9 | 0.50000 | 0.80902 | 0.91898 |
| 10 | 0.50000 | 0.50000 | 0.29782 |
| 11 | 0.50000 | 0.80902 | 0.83650 |
| 12 | 0.50000 | 0.50000 | 0.54707 |
| 13 | 0.50000 | 0.80902 | 0.99114 |
| 14 | 0.50000 | 0.50000 | 0.03514 |
| 15 | 0.50000 | 0.80902 | 0.13561 |

| $t$ | $a = 2$ | $a = 1+\sqrt{5}$ | $a = 4$ |
|---|---|---|---|
| 0 | 0.35000 | 0.35000 | 0.40001 |
| 1 | 0.45500 | 0.73621 | 0.96001 |
| 2 | 0.49595 | 0.62847 | 0.15357 |
| 3 | 0.49997 | 0.75561 | 0.51995 |
| 4 | 0.50000 | 0.59758 | 0.99841 |
| 5 | 0.50000 | 0.77820 | 0.00636 |
| 6 | 0.50000 | 0.55856 | 0.02526 |
| 7 | 0.50000 | 0.79792 | 0.09850 |
| 8 | 0.50000 | 0.52180 | 0.35518 |
| 9 | 0.50000 | 0.80728 | 0.91610 |
| 10 | 0.50000 | 0.50307 | 0.30743 |
| 11 | 0.50000 | 0.80899 | 0.85167 |
| 12 | 0.50000 | 0.50006 | 0.50531 |
| 13 | 0.50000 | 0.80902 | 0.99989 |
| 14 | 0.50000 | 0.50000 | 0.00045 |
| 15 | 0.50000 | 0.80902 | 0.00180 |

*Results from Franco Vivaldi, "An Experiment with Mathematics," in* Exploring Chaos, *ed. Nina Hall (1991), 33, 39.*

What is remarkable about the second table is that for the two nonchaotic systems, after a relativity brief variation (3 cycles for the steady system and 12 cycles for the oscillating system), the population pattern returned to exactly what it had been with the corresponding population in the first chart and, once returned, stays "corrected" forever. By contrast, for the chaotic system, by the end of the 15th cycle, there is absolutely no similarity between the first and second table.

Thus, small errors are forgiven in nonchaotic systems; they "come out in the wash." For a chaotic system, however, even the tiniest error completely changes the results. This is basically a generalization of the previously discussed phenomenon of sensitive dependence on initial conditions. The patterns in a chaotic system are completely altered not only by changes in initial conditions but by any change along the way. This is sometimes called the butterfly effect, so named because as small an act as a butterfly flapping its wings can change the weather around the world (although, of course, we never know in exactly what way).

You cannot rely on extra precision and carefulness to solve this problem. If you were 100,000 times more accurate in the second table, you merely would delay the onset of total variation twice as long.

The second lesson of chaos, then, is that for chaotic systems long-range prediction is impossible. Absent perfect, superhuman understanding of conditions, the variations caused by the tiniest error increase dramatically as time goes on. As Franco Vivaldi explains,

> *We have arrived at the core of the issue, the realization that there are systems, even within mathematics, that are both deterministic and unpredictable. We cannot blame this failure on the influence of unknown factors, because there are none. It is rather the result*

*of our own terminal inability to measure or represent the present with infinite precision.*

But all is not lost. Even though precise prediction is impossible, ignorance is not endemic. Chaotic systems are unpredictable, yet they may still be bounded. The precise orbits of some planets may be chaotic, but they do not crash into one another. Moving back to Earth, we cannot predict the weather precisely even a few days into the future, but we can tell that it will not snow in Disney World on Labor Day.

In chaos, there is also a tendency for systems to gravitate toward various areas, which are known as strange attractors. While the precise direction in which any given point will leave a strange attractor is unpredictable, it tends to return to the general area, sooner or later.

= = =

There is another interesting aspect of chaos, which sometimes goes by the name of complexity. This requires an examination of the transition from stability to chaos. In the second of our three models, we witnessed an oscillation between two points. This is an example of what is called bifurcation. But as the growth parameter *a* increases toward the point of chaos (3.57), the oscillation goes from 2 to 4 points of variance, and then to 8, 16, and so on, to infinity, with the bifurcations occurring at shorter and shorter intervals.

This area of rapidly increasing bifurcation has been termed the edge of chaos. It combines features of stability (there are, after all, recognizable patterns) with attributes of chaos (rapid, sudden changes, for example). Some believe that systems, be they biological or computer, are capable of the most complex, sophisticated adaptations, be they evolutionary or computational, if they are at the edge of chaos, neither too stable nor too chaotic.

The ability to adapt, of course, is essential for survival in a changing environment. But mere alteration is not enough. In order for complex systems to thrive, they also must maintain coherence under change. This combination—adaptation and coherence—has been identified in a host of what John Holland terms complex adaptive systems, from the human immune system to a rain forest ecosystem. The most successful systems may be those that find the region between ordered and chaotic states. The ideal state, in other words, should contain "just the right balance of stability and fluidity."

This state has been analogized to the phase transition that water undergoes as temperatures vary: chaos is like a gas, disorganized and frenzied, and order is like ice, rigid and solid. The ideal mix combines features of both, because neither status by itself is conducive to successful adaptation: "Highly chaotic networks would be so disordered that control of complex behavior would be hard to maintain. Highly ordered networks are too frozen to coordinate complex behavior."

≥ ≥ ≥

A different mathematical image is created by those systems which are generally stable systems but possess a few points where extreme change is possible. The mathematical model for this system comes from what is known as catastrophe theory. Metaphorically, this is the case of the straw that breaks the camel's back. There is either no change or only minimal change for a period of time, but then a small move results in an enormous disruption. Mathematically, the small change transforms a system from a stable attractor to an unstable attractor, and then a discontinuous shift moves the system to a new stable attractor.

A graph of a professor's grading system might reflect catastrophe (which is certainly what my students term mine). If limited to grading choices of A, B, C, or F, for example, the increase in a test score from 80 to 89 may have no effect on a course grade, but increasing just one point, from an 89 to a 90, catapults the student onto a whole new level.

Inordinate hoopla surrounded catastrophe theory when it was first developed. It was advertised as a virtual theory of everything—science, politics, and society. As might be expected, the claims that a mathematical theory could *predict* political or social upheavals largely have been discredited. Despite its limitations, however, catastrophe theory provides interesting *illustrations,* by creating models of systems that only occasionally display spasms of sudden change.

÷ ÷ ÷

The geometric configuration known as a fractal is another important explanatory tool. This is a shape that, unlike its Euclidean counterpart, not only is irregular but has the same degree of irregularity no matter how closely you look. If you take a close-up photograph of a part of an octagon, for example, you will get a simple straight-line segment, and even a circle will look

*Figure from Benoit B. Mandelbrot,* The Fractal Geometry of Nature *(1982), 51.*

flat on a small enough scale. But if you take an extremely close-up picture of a fractal, you will get a sharply jagged piece, with the same basic irregularity as before.

The self-similar pattern of fractals mirrors many natural objects—think of a stalk of broccoli. As the founder of the study of fractals, Benoit Mandelbrot, says:

> *Clouds are not spheres, mountains are not cones, coastlines are not circles, and bark is not smooth, nor does lightning travel in a straight line. All of these natural structures have irregular shapes that are self-similar. In other words, we discovered that successively magnifying a part of the whole reveals a further structure that is nearly a copy of the original we started with.*

However, just as Earth is not a perfect sphere, natural objects are not perfect fractals. While a tree does have branches with smaller branches down to the twig level, it does not split indefinitely into smaller and smaller versions of itself. A fractal is essentially a mathematical ideal, and hence an imperfect, albeit much-improved, model for the real world.

As mathematician Rudy Rucker explains, "Speaking casually, I say that a shape is 'fractal' if it has similar-looking structures on several different size scales—a line branches into lines that branch; a bump covered with bumps that are bumpy; a glob made of globs made of globs of globs. The more levels there are, and the more the levels look like each other, the more perfect a fractal I have."

## Is the Constitution Chaotic?

How meaningful is it to say that the Constitution is chaotic? Obviously, the mathematical theory of chaos was unknown when the Constitution was drafted. Moreover, the unquantifiable nature of constitutional doctrine and structure make formal modeling impossible. Finally, even nonlinear systems are not automatically chaotic.

Nonetheless, even absent a precise analogue, the theory of chaos offers rich, new ways of thinking about how systems behave over time. It certainly is plausible that constitutional jurisprudence, a discipline that considers whether rulings over time are consistent with "a line of cases" or "a pattern of decisions," will benefit from asking whether we have something akin to a straight line or a fractal and whether there is a chaotic or a predictable pattern. Indeed, the very structure of both the document and the government system it created leads to patterns that are reminiscent of those in modern chaos theory.

For example, our federalist system can be seen as a kind of fractal structure. A picture of the governing design for the nation would reveal, rather than a simple government structure, the self-similar pattern associated with fractals. From the largest, federal level of government, we split off into 50 small state governments, and then to the thousands of smaller county and city governing bodies. At each level we find elected officials, passing laws or hiring workers, raising taxes and spending money.

But as with fractal approximations of nature, one should not expect a perfect pattern of replication: "The drawback in using a fractal curve as a coastline model is that the fractal has too much detail. . . ." Similarly, the local town council has both a much less intricate structure than and different responsibilities from the federal government. Again, as with fractals, the interrelated similarity is noteworthy, although incomplete.

+ + +

On a more fundamental level, the entire issue of constitutional interpretation by the Supreme Court has much in common with a dynamic, chaotic system. Most importantly, the mathematical concept of iteration, or feedback, is paralleled by the Court's use of its own precedent to decide future cases.

Under the doctrine of *stare decisis,* each decision builds on previous ones. Just as the population density at one time depends on the density at a preceding time, each ruling depends on how the Court ruled the previous time a similar case was decided. Moreover, with both population and precedent, the longer the time span under consideration, the more cycles of iteration occur and the greater the likelihood of complexity.

This can be seen in lines of cases championed by all sides of the ideological divide. The 1922 case granting a Fifth Amendment right to compensation for a regulation making it "commercially impracticable to mine certain coal" led to the 1992 case requiring compensation for a law preventing an owner from any construction on beachfront property, which led to a 1994 ruling that a city could not require a store to dedicate part of its property as a public bicycle path in exchange for the granting of a building permit.

Similarly, the Court's striking down of a law barring parents from sending their children to private school led to its declaration that bans on the sale of contraceptive devices to married couples were unconstitutional, which, in turn, led to *Roe v. Wade*.

Even though the Constitution in general and the Bill of Rights in particular are relatively short and simple, the application of their principles over several centuries would be expected to become increasing complicated. In constitutional law, as in mathematics, "How can simple rules lead to complex phenomena? Via long runtime."

Chaos sheds, appropriately enough, a complex light on the competing theories of constitutional interpretation. On the one hand, the fact that the current state of constitutional doctrine is vastly different from what the framers would have envisioned should not be considered surprising.

Even those who believe that constitutional decisions should be based solely on the original intent or understanding of the framers recognize that it is not always easy to ascertain the unified intent of a group of people who lived more than two centuries ago. As Professor Richard Kay has noted, a judge following original intent need not divine the framers' intent precisely to decide whether a particular government action is constitutional, "because the alternatives are binary. The question can and can only be answered 'yes' or 'no,' and since the judge must give some answer, it follows that he need not answer with certainty. All he needs to do is decide which of the two possible answers in that case is *more likely* correct."

Over time, as implied by the theory of chaos, this lack of precision will result in a very different constitutional path from what the framers ever would have expected. The inevitability of error limits how close to original intent anyone can expect modern constitutional rules to be. As we have seen with chaotic systems, "the smallest change leads ultimately to quite different detailed development."

Accordingly, the following conclusion about the pattern of constitutional development, by constitutional scholar Paul Brest, is very close to what chaos theory would predict:

> *[I]f you consider the evolution of doctrines in just about any extensively-adjudicated area of constitutional law—whether 'under' the commerce, free speech, due process, or equal protection clauses—explicit reliance on originalist sources has played a very small role compared to the elaboration of the Court's own precedents. It is rather like having a remote ancestor who came over on the Mayflower.*

It is unrealistic to expect that our current doctrine would fulfill the framers' expectations. Even were society not to change, the long-term iteration of inevitably imperfect decisions would tend to lead to results that were unforeseeable initially. And errors in constitutional decision making can have enormous long-term repercussions.

This observation predates the Constitution, as evidenced by Jonathan Swift's description in *Gulliver's Travels:*

> *It is a maxim among these lawyers, that whatever hath been done before, may legally be done again; and therefore they take special care to record all the decisions formerly made against common justice, and the general reason of mankind. These, under the name of precedents, they produce as authorities, to justify the most iniquitous opinions, and the judges never fail of directing accordingly.*

To that extent, Chief Justice Rehnquist surely is correct when he says that it is the Court's "duty to reconsider constitutional interpretations that 'depar[t] from a proper understanding' of the Constitution. . . ." Of course, that reconsideration does not take place in a vacuum. Even erroneous decisions might sometimes be better left untouched. As the Supreme Court said when declining to overrule *Roe v. Wade,* precedent should not be reversed if it would cause "serious inequity to those who have relied upon it" or impair the "public faith in the judiciary as a source of impersonal and reasoned judgment."

× × ×

The issue of predictability, and its real-world counterpart, reliability, is a two-edged sword. Imagine the Supreme Court's decision making as a form of iterative decisions; under *stare decisis* each new case builds on the holding that preceded it. A general function would look like $X_{t+1} = f(X_t)$, where $X_t$ is the holding of the preceding case and $f(X_t)$ is the application of the holding from the preceding case to the facts of the next case, which leads to $X_{t+1}$, the holding in the next case.

Two kinds of instability become apparent. If we ignore precedent, fail to use $X_t$ at all, then the result, $X_{t+1}$, of the next case obviously is unpredictable. However, if we continually build on an erroneous decision—the original $X_t$ was wrong—we may end up with errors of increasing magnitude.

We are left with an inevitable tension. In the short term, if courts do not follow precedent reliably, the law itself is unreliable. Over the long term, however, while weighing the virtues of *stare decisis,* it is appropriate also to remember that, "in a chaotic system errors grow at an accelerating rate."

≠ ≠ ≠

Sometimes constitutional change doesn't grow; it happens all at once. Catastrophe theory may present a picture of what commonly is called a "constitutional moment." At various points in our history, cataclysmic outside events have resulted in a major shift in constitutional interpretation. The Great Depression, for example, witnessed a radically new view of the constitutionality of economic regulation. In the words of Professor Laurence Tribe, "The 1937 watershed marked a change in how courts interpreted [liberty]." According to Justice Anthony Kennedy, this transformation, "forecloses us from reverting to an understanding of commerce that would serve only an 18th-century economy. . . . Congress can regulate in the commercial sphere on the assumption that we have a single market and a unified purpose to build a stable national economy."

The revolution in constitutional interpretation can be seen as analogous to the jump discontinuity illustrated in catastrophe theory, when a model shifts to an entirely new pattern. Metaphorically, the Depression's upheaval caused the restrictive view of the government's power over economic matters to be the equivalent of an unstable attractor, and the Court responded with a discontinuous shift to what has turned out to be a far more stable, even if not a stationary, situation.

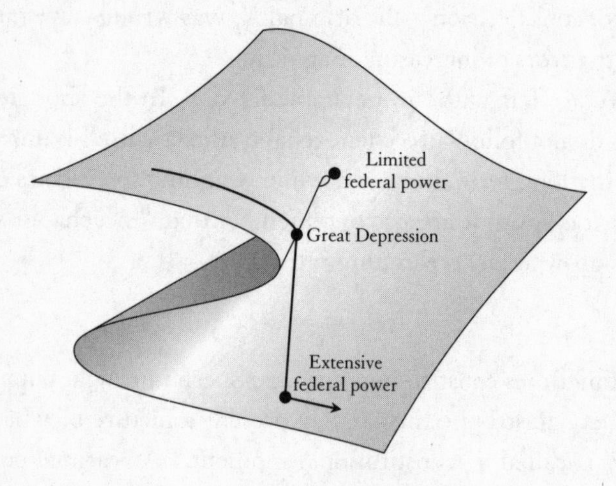


*Figure adapted from John L. Casti,* Complexification *(1994), 59.*

× × ×

Despite chaos and catastrophe, systems do not necessarily careen into oblivion. Within the seemingly random action of many chaotic systems, for example, one can find security in the fact that boundaries still exist. The attractors keep the system from racing off in one direction indefinitely.

Our system of constitutional jurisprudence may be seen as similarly bounded. One of the main restraints is the text of the document itself. As Dean Frederick Schauer has noted, the language used in the Constitution can act as the boundary restricting, if not determining, judicial decisions: "An interpretation is legitimate (which is not the same as correct) only insofar as it purports to interpret some language of the document, and only insofar as the interpretation is within the boundaries at least suggested by the language."

Without doubt, constitutional language can be stretched, but there are institutional constraints. As Justice White warned: "The Court is most vulnerable and comes nearest to illegitimacy when it deals with judge-made constitutional law having little or no cognizable roots in the language or design of the Constitution."

= = =

Other boundaries limiting erroneous decisions are created by the system of appointment laid out in the Constitution. The Court is replenished, slowly and irregularly, but replenished all the same, by presidential appointment with Senate confirmation. Historically, the political appointment process has served as its own kind of strange attractor, eventually tugging the Court until it veers off in a new direction.

This repeated change of direction is characteristic of chaotic systems. Such systems are considered topologically transitive, meaning that they run all over the map sooner or later. Similarly, one should expect both so-called conservative and liberal rulings to keep popping up, although sometimes the wait may seem excruciating.

A substantially simpler mathematical model of this phenomenon is the helix. Famous as the heart of the structure of DNA, a helix resembles an endless Slinky, or a curve that twists around a cylinder (see figure on following page).

Mathematician Rudy Rucker has pointed out that a helix creates a powerful metaphoric image, since it combines circular motion (as on a flat plane) with upward movement (as soaring through space):

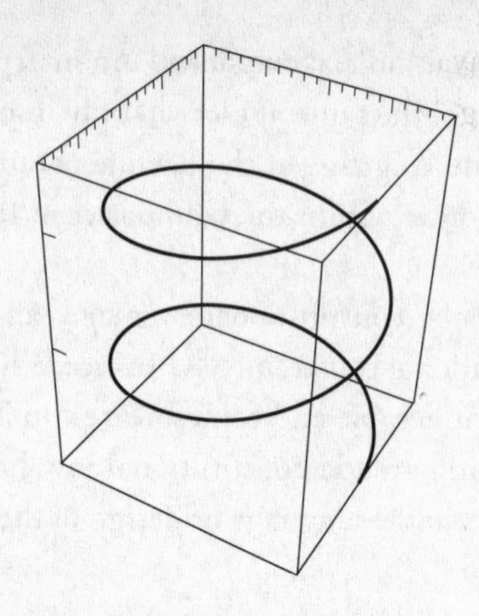

*Human intellectual development can be thought of as helical. Society is always circling back and forth between liberalism and conservatism, yet the conservatism of the 1980's is not quite the same as the conservatism of the 1950's. Yesterday's cycle through liberalism informs today's conservatism, just as today's conservatism will inform tomorrow's liberalism. The motion is helical, rather than circular. On one level we might say that we are back where we started, but this is not the case. The past is not forgotten.*

≥ ≥ ≥

In addition to our constitutional jurisprudence, the nation's constitutional political structure also can be illuminated by chaos theory. The inherent tensions within and the constant perturbations from outside ensure that our system continues to be both unpredictable and dynamic. To survive, the political system has had to display many of the features associated with classical complex adaptive systems. For more than two centuries, it has managed to adapt repeatedly to an extraordinary array of small alterations and grand upheavals, both external and internal, while at the same time maintaining "coherence under change." It is not surprising, then, to see built into our Constitution the defining feature of other complex adaptive systems, a "balance of stability and fluidity."

The predominant agent of change in a democracy is the ballot box. When the public decides that it no longer approves of the work of its political representatives, it can replace them in the next election. The people entrust political power to their representatives and, in the words of George Washington, "whenever it is executed contrary to their Interest, their servants can, and undoubtedly will be, recalled." But the framers saw that too rapid a change could threaten liberty as much as a structure that was too resistant to change.

The constitutional structure of the new government had to be built in recognition of these contradictory dangers. Thus, there was widespread support for the initial proposition that "[f]requent elections are unquestionably the only policy" that ensures that representatives are both sympathetic to the needs, and dependent on the approval, of their constituents. However, the enthusiasm for this position was tempered by the conviction that not all change, even that which is desired by the majority, is good:

> *The mutability in the public councils arising from a rapid succession of new members, however qualified they may be, points out, in the strongest manner, the necessity of some stable institution in the government. . . . From this change of men must proceed a change of opinions; and from a change of opinions, a change of measures. But a continual change even of good measures is inconsistent with every rule of prudence and every prospect of success.*

Thus, an innovative structure was created to balance these concerns. The members of the two houses of the legislature were given differing terms of office (2 years in the House of Representatives and 6 years in the Senate), with the Senate divided so that only one-third of its members could be changed in any single election.

The mix of alterability and stability can be seen if we imagine that the overwhelming majority of Americans suddenly gets swept up in a desire for a radical overhaul of the government. Unlike in a monarchy, the people are free to effect a change. Yet under the constitutional plan, they must proceed in a piecemeal fashion. At the first opportunity, they can replace all of the House of Representatives but only one-third of the Senate. Thus, it takes two elections for a majority in both Houses to be replaced. After the first election, the beginning of a political agenda will be visible, but major revisions in law could be prevented and proposals for constitutional change that would solidify the position of those in office could be defeated. Within two years, the fervor of an illegitimate movement may well have died out. A genuine popular movement, however, will be able to continue its electoral success and obtain, after the second election, a majority in both Houses.

Thus, the framers attempted to create the necessary balance between innovation and consistency. This need to preserve coherence under change is exemplified by the Supreme Court's decision striking down mandatory term limits, *U.S. Term Limits, Inc. v. Thornton.* In *Thornton,* Arkansas had attempted to limit its federal representatives to three terms in the House and two in the Senate. The motivation behind these limitations was to remove the "entrenched incumbency." While this goal is consistent with the need for fluidity, term limits may be too effective. They threaten the stability that has coexisted with the fluidity. As Roger Sherman, a delegate from Connecticut to the Constitutional Convention, stated, the ability to reelect effective legislators "as often as the electors shall think fit . . . will give greater stability and energy to government than an exclusion by rotation. . . ."

÷ ÷ ÷

This mix of stability and energy was also seen at the edge of chaos. The systems that are most successful at adapting to change in their environments combine elements of both order and chaos.

The Supreme Court itself used a similar metaphor, describing the appropriate regulation of election law as one that "in no way freezes the status quo, but implicitly recognizes the potential fluidity of American political life." Thus, the Court ruled that a limitation on the ability of independent candidates and new political parties to obtain access to the ballot eliminated too much fluidity to withstand constitutional challenge.

In *Williams v. Rhodes,* for example, the Supreme Court struck down a series of requirements that "made it virtually impossible for a new political party to be placed on the ballot." The state tried to justify its requirements on the theory that a state "may validly promote a two-party system in order to encourage compromise and political stability." This was rejected by the Court:

> *Competition in ideas and governmental policies is at the core of our electoral process and of the First Amendment freedoms. New parties struggling for their place must have the time and opportunity to organize in order to meet reasonable requirements for ballot position, just as the old parties have had in the past.*

In *Timmons v. Twin Cities Area New Party,* however, the Court upheld a Minnesota law banning fusion candidacies, that is, a candidate's appearance on the ballot as the representative of more than one party. The law was justified on numerous grounds, including the concern that a minor party might gain votes from people interested in voting for a candidate already on the Democratic or Republican ticket. This would permit the minor party to bootstrap its way to major party status and circumvent the normal nominating petition requirement for minor parties. The Court also stated, however, that states had "a strong interest in the stability of their political systems" and that a state legislature may "decide that political stability is best served through a healthy two-party system."

This last point, if carried too far, could create a real danger that, in the name of political stability, legislators would try to grant "the two old, established parties a decided advantage over any new parties struggling for existence." A valid interest in political stability should not be at the expense of the natural fluidity of U. S. politics.

+ + +

A quantitative analysis of the Constitution is impossible to perform and meaningless to imagine. There is no model to predict the next Supreme Court ruling or the outcome of the next election. Nonetheless, there is much that those studying the Constitution can learn from chaos.

Most fundamentally, "it is plain . . . that realistic accounts of social and political processes are apt to be nonlinear." There is a grand flexibility for any thinking, especially concerning the Constitution, that comes from realizing that big changes can follow small events, that repetition over time and multilateral relationships can create incredibly complex results, and that even the smartest person cannot always predict what will happen next. In the words of the biologist and chaos pioneer Robert May, "Not only in research, but also in the everyday world of politics and economics, we would all be better off if more people realized that simple nonlinear systems do not necessarily possess simple dynamical properties."

A strong argument can be made that the framers designed a Euclidean Constitution. It turns out that they may have created a chaotic Constitution as well.

# 11

# The Mathematics of Limits

**But a constitution is framed for ages to come, and is designed to approach immortality as nearly as human institutions can approach it.**

***—Chief Justice John Marshall***
**Cohens v. Virginia *(1821)***

David Berlinski, in *A Tour of the Calculus,* tells the story of what happened when he was giving a mathematics lecture in Prague. He began by describing a function: "A function indicates a relationship in progress, arguments going to values. Given any real number, the function $f(x) = x^2$ returns its square, *tak?*"

He continued, "The image of a machine, something like a device making sausages, is irresistible. *In* go the arguments 1, 2, 3, *out* come the values 1, 4, 9."

He then asked, "What happens to the values of $f$ as its arguments approach 3? They approach, those values, the number 9, so that the function is now seen as running up against a *limit,* a boundary beyond which it does not go."

He then made a statement that he himself found astonishing: "It is when functions are seen in *this* context that the poignancy of the process becomes for the first time palpable." In the example of $f(x) = x^2$, the function achieves a moment of blessed release at the number 3; *there* $f(3)$ *is* 9, the process of getting closer and closer over and done with. The limit is reached.

Next he wrote on the board $f(x) = (x^2 - 1)/(x - 1)$. "Here," he said, "is another story. As $x$ approaches ever more closely the number 1, $f(x)$ gets closer and closer to the number 2. It *approaches* 2 as a limit. But at 1, $f(x)$ lapses into nothingness *because* at 1, $(x^2 - 1)/(x - 1) = (1 - 1)/(1 - 1)$ which is simply (0/0). Because we cannot divide by zero, at its limit, this function is *undefined.* The function gets closer and closer to its limit, but, you see, *it never reaches* that limit."

"*Tak,*" says someone in the audience, "*like man to God.*"

≠ ≠ ≠

Speed is arithmetic. You take the distance you travel, divide it by how long you travel, and the quotient arrived at is speed (rate of speed = distance/time). If I travel 100 miles from 11 A.M. till 1 P.M. (for two hours), my average speed is 50 (100/2) miles per hour. But perhaps my speed varied; maybe I traveled at 60 miles per hour for the first hour and only 40 miles per hour for the second. My average speed would still have been 50 miles per hour for the whole trip, but you could not know what my speed was at exactly noon.

In fact, there is no way to determine precisely one's speed at a specific moment. An instant of time has no duration; it is no time at all. Try to plug it into our speed formula, and you are stuck: you can't divide by 0. We can divide distances by 1 hour, 1 minute, 1 second, but not by one instant.

To deal with this problem, mathematicians created the concept of limits. The trick is to take smaller and smaller time intervals around noon and measure the average speed as we get closer and closer to noon. If the average speed for smaller intervals, 2 seconds, 1 second, ½ second, and so on, goes from 54½, to 54¼, to 54⅛, and so on, we could say that the average speed is approaching 54. Unlike normal speed calculations, instantaneous speed "is *not* defined as the quotient of distance divided by time. Rather we have introduced the idea of taking a number that is *approached* by average speeds."

Limits also are critically important for the understanding of functions, as was shown in the story at the beginning of this chapter. That story told of the function $f(x) = (x^2 - 1)/(x - 1)$. This function is not defined when $x = 1$, because that would reduce to $(1 - 1)/(1 - 1)$, or 0/0, which involves impermissible division by 0. But $x$ can get as close as you want to 1, and as $x$ *approaches* 1, the function $f(x)$ *approaches* 2.

A limit in mathematics has a very rigorous and formidable definition. What the definition is trying to communicate, though, is simply "our intuition of something tending toward or approaching an ultimate value. . . ."

One of the reasons that the concept of limits has such power and utility is that it permits mathematical understanding to be based on what something is tending toward, even though it may never actually get all the way there. We can discuss a function as approaching a limit despite the fact that it never actually reaches that point:

> *The* concept *of a limit is simple. It is the* definition *that is complex. The concept involves nothing more obscure than the idea of getting closer and closer to something. It suggests the attempt by one human being to approach another, and the inexpungeable thing in love as in mathematics is that however the distance*

*decreases, it often remains what it always was, which is to say, hopelessly poignant because hopelessly infinite.*

× × ×

There is the same poignancy within the Constitution as there is with the mathematical concept of limits. Nowhere is this better exemplified than in the Preamble to the Constitution, which promises "a more perfect Union." How can you be more perfect? Was the Union under the Articles of Confederation perfect?

To answer the second question first, the general understanding was that things were a mess. Justice Joseph Story, in his *Commentaries on the Constitution of the United States,* spent the entire ten-page discussion of the phrase "more perfect Union" focusing on the weaknesses faced by the country before the Constitution—economically, politically, and in its dealings with foreign countries.

The Supreme Court similarly has interpreted the phrase "more perfect Union" to indicate that the framers intended to strengthen the powers of the federal government over the ineffectual regime under the Articles of Confederation:

> *All acknowledge that they were convened for the purpose of strengthening the confederation by enlarging the powers of the government, and by giving efficacy to those which it before possessed, but could not exercise. They inform us themselves, in the instrument they presented to the American public, that one of its objects was to form a more perfect union. Under such circumstances, we certainly should not expect to find, in that instrument, a diminution of the powers of the actual government.*

So, "more perfect Union" is, in a sense, saying "a better Union." But it is more than that. What "more perfect" really stresses is not merely improvement over the status quo but changing so as to be closer to perfect. The goal of a perfect system, a perfect democracy, is an ideal, but an ideal that never can be attained. As with the limits of a function, we can ever improve, ever progress toward our goals, but those goals may prove forever unreachable.

= = =

The very system of representative democracy reflects a realization that a pure democracy, in which all the citizens gather and decide every issue, is impossible. Our system is premised, however, on the concept that while such a system "would be unable to govern a nation . . . it would remain an *ideal to be approximated* as nearly as possible in practice."

The recognition of the limits of the Constitution also can be seen in James Madison's discussion of the dangers of insurrection. He argued that a strong national government would prevent local uprisings such as Shays's Rebellion but conceded that the federal government could not prevent a nationwide rebellion: "[I]t is a sufficient recommendation of the Federal Constitution that it diminishes the risk of a calamity, for which no possible constitution can provide a cure."

Two quite different theories that deal with the intersection of fundamental constitutional rights and group hatred also imply the mathematical concept of limits. The first involves government suppression of "hate speech," communication that vilifies individuals or groups on the basis of characteristics such as race, religion, and gender. A law banning cross burning was struck down by the Supreme Court, on the theory that the regulation was not "reasonably necessary" to ensure the "basic human rights of members of groups that have historically been subjected to discrimination." The constitutional way for a government to deal with noxious speech that threatens others is by education and example. As Justice Louis Brandeis wrote, "the remedy to be applied is more speech not enforced silence."

The conviction that good speech can drive out harmful speech dates at least to the seventeenth century writing of John Milton:

> *And though all the winds of doctrine were let loose to play upon the earth, so Truth be in the field, we do injuriously by licensing and prohibiting to misdoubt her strength. Let her and Falsehood grapple, who ever knew Truth put to the worse, in a free and open encounter?*

Much of the heart of the First Amendment can be seen in the faith that, over time, falsity will be exposed to all. The limit, if you will, of a free and open debate, is the truth.

≥ ≥ ≥

Sometimes, it takes a long while for a series to get even reasonably close to its limit. Consider, for example, the series $1 - 1/3 + 1/5 - 1/7 + \ldots$, which eventually approaches $\pi/4$. This series is actually a very slow way to calculate the value of $\pi$. It would take some 628 terms to approximate $\pi$ to just two decimal places (3.14). By contrast, the series $1/1^4 + 1/2^4 + 1/3^4 + 1/4^4 + \ldots = \pi^4/90$ converges much more quickly, reaching a better approximation of $\pi$ with fewer than ten terms. Put simply, it takes much longer for some series to approach their limit than others.

The same applies to speech. Not all corrections occur in a brief interval after the first rebuttal. As Professor Harry Wellington noted,

> *In the long run, true ideas do tend to drive out false ones. The problem is that the short run may be very long. . . . [T]ruth may win, and in the long run it may almost always win, but millions*

*of Jews were deliberately and systematically murdered in a very short period of time. [B]efore those murders occurred, many individuals must have come to have false beliefs.*

The reason for allowing the hateful speech, then, is not that we know it will be corrected before harm is done. Education often will be the best remedy, but the protection of the First Amendment serves another purpose. If the government were permitted to make content-based distinctions on speech, it would raise the "realistic possibility that official suppression of ideas is afoot."

÷ ÷ ÷

For some issues, however, it is unacceptable to sit and wait for the right result. Some have argued that antidiscrimination laws, particularly in employment, are unnecessary, because over time the bigots will be harmed economically. As Judge Richard Posner explains, if those with racial animus forgo economically beneficial transactions with African Americans, "[t]he least prejudicial sellers will come to dominate the market" and bigotry will "disappear eventually from competitive markets."

There is some debate over whether unregulated economic forces actually would be able to eradicate racial discrimination in the marketplace. Perhaps the limit of an unregulated marketplace is indeed the end of irrational employment discrimination along racial lines. But this is one of those areas where mathematics misses the most important variable: the pain and indignity caused by discrimination while we wait for that beneficent limit to be approached.

The most eloquent expression of this concept can be found in Reverend Martin Luther King Jr.'s historic "Letter from a Birmingham Jail." Written in 1963, after he was arrested during a civil rights protest, his letter to eight clergymen explains why "justice too long delayed is justice denied":

> *But when you have seen vicious mobs lynch your mothers and fathers at will and drown your sisters and brothers at whim; when you have seen hate-filled policemen curse, kick and even kill your black brothers and sisters; when you see the vast majority of your twenty million Negro brothers smothering in an airtight cage of poverty in the midst of an affluent society; when you suddenly find your tongue twisted and your speech stammering as you seek to explain to your six-year-old daughter why she can't go to the public amusement park that has just been advertised on television, and see tears welling up in her eyes when she is told that Funtown is closed to colored children, and see ominous clouds of inferiority beginning to form in her little mental sky, and see her beginning to distort her personality by developing an unconscious bitterness toward white people . . . when you are forever fighting a degenerating sense of "nobodiness"—then you will understand why we find it difficult to wait.*

12

# The Limits of Mathematics

**As far as the laws of mathematics refer to reality, they are not certain, and as far as they are certain, they do not refer to reality.**

***—Albert Einstein,* Sidelights on Relativity *(1923)***

## School Days

"Sam, if a man can walk 3 miles in 1 hour, how many miles can he walk in 4 hours?"

"It would depend on how tired he got after the first hour," replied Sam.

The other pupils roared. Miss Snug rapped for order.

"Sam is quite right," she said. "I never looked at the problem that way before. I always supposed the man could walk 12 miles in 4 hours, but Sam may be right: that man may not feel so spunky after the first hour. He may drag his feet. He may slow up."

Albert Bigelow raised his hand. "My father knew a man who tried to walk 12 miles, and he died of heart failure," said Albert.

"Goodness!" said the teacher. "I suppose *that* could happen, too."

"Anything can happen in 4 hours," said Sam. "A man might develop a blister on his heel. Or he might find some berries growing along the road and stop to pick them. That would slow him up even if he wasn't tired or didn't have a blister."

"It would indeed," agreed the teacher. "Now, if you are feeding a baby from a bottle, and you give the baby 8 ounces of milk in one feeding, how many ounces of milk would the baby drink in *two* feedings?"

Lisa Staples raised her hand. "About 15 ounces," she said.

"Why is that?" asked Miss Snug. "Why wouldn't the baby drink 16 ounces?"

"Because he spills a little each time," said Lisa. "It runs out of the corners of his mouth and gets on his mother's apron."

By this time the class was howling so loudly the arithmetic lesson had to be abandoned. But everyone had learned how careful you have to be when dealing with figures.

\+ + +

E. B. White, who recounted the above tale in *The Trumpet of the Swan,* was right: One really does have to be careful when dealing with figures. There is so much that constitutional lawyers can learn from mathematics, yet at the same time there is so much that can mislead. Mathematical ideas and models can help to clarify our thinking, but not if we lose sight of their limitations.

It is foolhardy, in the words of Douglas Hofstadter, to study mathematics "wishing to find laws for cloud motion and hoping to get away with treating clouds as stable, solid, sharp-edged objects—that is hoping not to be forced to take into account the fact that clouds are tenuous, amorphous, boundary-less puffs of fluff made out of molecules rushing all about every which way."

Constitutional law is similarly tenuous and amorphous: the issues being dealt with are political, subjective, and dependent on an enormous array of factors. The reality of constitutional jurisprudence is "infinitely complex and impossible to capture completely in any mathematical model." Accordingly, we should not look to mathematics for authoritative answers to constitutional questions.

Moreover, we must be careful when we do use mathematics not to confuse the precision of a number with the accuracy of a result. My father, a retired accountant, tells me that when he gave people estimates he always rounded to the nearest dollar. He says that no matter how often he told his clients that the figures were mere approximations subject to substantial revision, they couldn't help but view a sum given to the exact penny as truth.

This phenomenon also has been described by mathematician and author David Berlinski:

> *Quantitative work has its special dangers in that spurious moral grandeur is generally attached to any formulation computed to a large number of decimal places. Faced with an authoritative avowal that pig-iron production is increasing at an annual rate of 13.295674 percent, only the incorrigible skeptic will raise the possibility that pig-iron production may not be increasing at all.*

Even if the numbers are correct, a related concern is that their concreteness will overwhelm factors that are not as quantifiable. It is simply inappropriate to present DNA evidence as showing with a 99.99 percent certainty that the defendant's blood was found at the murder scene, without establishing with comparable certainty that the bloodwork was done properly, without carelessness or deliberate misidentification. As the California Supreme Court cautioned in 1968, "Mathematics, a veritable sorcerer in our computerized society, while assisting the trier of fact in the search for truth, must not [be allowed to] cast a spell over him."

Moreover, many, if not most, important questions cannot be quantified. There is no way to calculate the probability that an individual is lying or the intensity of feelings such as anger or love.

Ultimately, the most important limitation of mathematics is that it is utterly incapable of making the sorts of judgments and interpretations that lie at the heart of the Constitution. Periodically, some mathematical minds have believed that the world could be "mathematized." Around 500 B.C., followers of Pythagoras created what amounted to a religion centered around mathematics. In addition to prohibiting the wearing of wool clothing and the eating of beans, they tried to live by their motto: "All is number." More than 2,000 years later, one of the co-inventors of calculus, Gottfried Leibniz, dreamed of creating a "calculus of reasoning," in which all ideas could be "factored" into their prime components, the fundamental concepts of mankind. Thus, people who had a disagreement could readily resolve their dispute by saying, "Let us calculate."

Those who claim such a magical power for mathematics justifiably lose credibility. For students of the Constitution, mathematics cannot pretend to be able to answer the hard questions of values and history, of inclusion and distinction. From abortion to affirmative action, from free speech to freedom of religion, mathematics can help to create an understanding that permits a greater sensitivity to cause and effect, as well as to the underlying assumptions that lead to opposing conclusions. But the choice between competing values emphatically is not a mathematical excursion. There is no calculus, no graph, no logical formula that can make the selection of fundamental principles and values. No computer can be programmed to render a constitutional decision.

For constitutional law, there is no "theory of everything," and if there were, it would not be mathematical. As Keith Devlin put it in *Mathematics: The Science of Patterns,* "Mathematics never captures all there is to know about anything. Understanding gained through mathematics is just a part of a much larger whole." Mathematics can illuminate various areas of our Constitution, but it is not the only source of light.

Suddenly, Gödel bolted from his chair and shouted, "On the contrary. I know how it could happen." Both Einstein and Morgenstern had to restrain Gödel so that he would return to his seat and stop his explanation. Finally, the exam resumed, and, shortly thereafter, Gödel became a citizen.

# Notes

Works frequently cited have been identified by the following short titles:

| | |
|---|---|
| *Debates in the Federal Convention* | James Madison, *The Debates in the Federal Convention of 1787 which Framed the Constitution of the United States of America.* (1835; reprint, Westport, Conn.: Greenwood Press, 1970), 239. |
| *Jefferson: Writings* | *Thomas Jefferson: Writings,* ed. Merril D. Peterson (New York: Viking Press, 1984). |
| Kaminski and Saldino, *Documentary History* | John Kaminski and Gaspare Saldino, eds., *The Documentary History of the Ratification of the Constitution* (Madison: State Historical Society of Wisconsin, 1983). |
| Kurland and Lerner, *Founders' Constitution* | Philip B. Kurland and Ralph Lerner, eds., *The Founders' Constitution* (Chicago: University of Chicago Press, 1987). |
| *Papers of Thomas Jefferson* | *The Papers of Thomas Jefferson,* ed. Charles Cullen (Princeton, N.J.: Princeton University Press, 1990) 370, 373. |
| Story, *Commentaries on the Constitution* | Joseph Story, *Commentaries on the Constitution of the United States* (1833; reprint, Durham, N.C.: Carolina Academic, 1987) 240. |

## Preface

12 *"Good—he did not have enough imagination":* Robert Osserman, *Poetry of the Universe* (New York: Doubleday, 1995), 89.

12 *"[t]he different branches of Arithmetic":* Lewis Carroll, *Alice's Adventures in Wonderland* (1865; reprint, Harmondsworth, Middlesex, England: Puffin Books, 1962), 127.

12 *an aesthetic quality to mathematics:* Edward Kasner and James Newman, *Mathematics and the Imagination* (New York: Simon & Schuster, 1940), 8, 466.

12 *"mathematics is the science of patterns":* Keith Devlin, *Mathematics: The Science of Patterns* (New York: Scientific American, 1997), 3. See also Ivars Peterson, *Islands of Truth: A Mathematical Mystery Cruise* (New York: Freeman, 1990), 31: "Mathematics is really the science of patterns. . . . Theories emerge as expressions of relationships among patterns—patterns of patterns."

12 *"If mathematics is anything":* Douglas Hofstadter, *Fluid Concepts and Creative Analogies* (New York: Basic Books, 1995), 70 (emphasis in original).

12 *"A page of history": New York Trust Co. v. Eisner,* 256 U.S. 345, 349 (1921).

12 *"All economical and practical wisdom":* Oliver Wendell Holmes Jr., *The Autocrat of the Breakfast-Table* (Boston: Phillips, Sampan, 1858).

13 "Doctrine of Fluxions": Letters from Thomas Jefferson to Robert Patterson, March 30, 1798, and March 31, 1798, in *Thomas Jefferson's Farm Book,* ed. Edwin Morris Betts (Charlottesville, Va.: University Press of Virginia, 1953), 51–52.

13 *"Jefferson's intellectual world":* I. Bernard Cohen, *Science and the Founding Fathers* (New York: Norton, 1995), 88.

13 *"[O]ne rebellion in 13 states":* Letter from Thomas Jefferson to James Madison, December 20, 1787, in *The Writings of Thomas Jefferson,* vol. 12, ed. A Lipscomb and Albert Berg (New York: Thomas Jefferson Memorial Association, 1903), 442.

13 *"given up newspapers":* Thomas Jefferson to John Adams, January 21, 1812, in *The Adams-Jefferson Letters: The Complete Correspondence Between Thomas Jefferson and Abigail and John Adams,* vol. 2, ed. Lester J. Cappon (Chapel Hill: University of North Carolina Press, 1959), 291.

13 *"deducing conclusions from axioms":* Morris Kline, *Mathematics in Western Culture* (1953; bk. ed., New York: Oxford University Press, 1956), 462–63.

14 *"said to have invented non-Euclidean legal thinking":* Jerome Frank, "Mr. Justice Holmes and Non-Euclidean Legal Thinking," *Cornell Law Quarterly* 17 (1931): 568, 572.

14 *"[a]xioms have been secularized":* Ibid., 576.

14 *"Any idiot can draw a graph":* David Berlinski, *On Systems Analysis* (Cambridge, Mass.: MIT Press, 1976), 75, quoting Nelson Norgood, *Sociological Sociometrics* (1969), 579.

15 *"people often charge that a knowledge":* John Allen Paulos, *Beyond Numeracy* (1991), 32.

## Introduction: The Ugliest Number in the Constitution

16 *"by adding to the whole Number":* U.S. Const., Art. 1, § 2, cl. 2.

16 *The idea of counting slaves:* For an excellent history of the development of the three-fifths clause, see John Kaminski, ed., *A Necessary Evil?* (Lunham, Md.: Rowman & Littlefield, 1995), 19–23.

17 *taxes would be apportioned:* Art. of Conf., Art. VIII.

17 *"in proportion to the whole number":* Kaminski, *A Necessary Evil?,* 238.

17 *"if an attempt should be made":* Report of the North Carolina Congressional Delegates to Governor Alexander Martin, March 24, 1783, in Kaminski, *A Necessary Evil?*, 20.

18 *"did not see on what principle":* James Wilson, July 11, 1787, in *Debates in the Federal Convention,* 239.

18 *"Are [slaves] admitted as Citizens?":* Ibid.

18 *"give such encouragement to the slave trade":* Gouverneur Robert Morris, July 11, 1787, in *Debates in the Federal Convention,* 240.

18 *"never confederate on any terms":* William Davie, July 12, 1787, in *Debates in the Federal Convention,* 240.

18 *"The principle of representation":* Story, *Commentaries on the Constitution,* 240.

19 *"What a strange and unnecessary accumulation":* Brutus III, *New York Journal,* November 15, 1787, in *Documentary History,* vol. 14, Kaminski and Saldino, 119–20. "Brutus" was the pseudonym of an anti-Federalist, but there is much uncertainty over his actual identity. See Kaminski and Saldino, *Documentary History,* vol. 13, 411–12.

19 *"Let the case of the slaves":* Federalist No. 54 (Hamilton or Madison).

19 *"Viewed in its proper light":* Story, *Commentaries on the Constitution,* 240–41.

20 *The South's share:* Donald G. Nieman, *Promises to Keep* (New York: Oxford University Press, 1991), 11.

20 *"The three-fifths compromise had given":* A. Leon Higginbotham, *The Black Journey: From Slavery to "Freedom" and Equality* (unpublished course notes, Library of Congress, 1993), 78–79.

## 1. Logic (Healthy and Ill)

23 *"It reminds me of an answer":* Charles L. Dodgson, *Euclid and His Modern Rivals* (2d ed. 1885; reprint, New York: Dover, 1973), 168.

23 *Teachings from the world of mathematics:* Compare Kline, *Mathematics in Western Culture,* 462–63 (stating that "the typical thought process of lawyers," as with the mathematician, involves, "deducing conclusions from axioms about undefined terms") with Laurence H. Tribe, "Taking Text and Structure Seriously: Reflections on Free-Form Method in Constitutional Interpretation," *Harvard Law Review* 108 (1995): 1221, 1224 n. 4 (stating that, "the 'Q.E.D.' that can properly end a mathematical proof knows no precise analogue in any branch of law"). See generally Kevin W. Saunders, "What Logic Can and Cannot Tell Us about Law," *Notre Dame Law Review* 73 (1998): 667; Kevin W. Saunders, "Informal Fallacies in Legal Argument, *South Carolina Law Review* 44 (1993): 343.

24 *"the most influential textbook":* Carl B. Boyer, *A History of Mathematics* (2d ed.; New York: John Wiley and Sons, 1991), 119.

24 *From a mere handful of axioms:* Kline, *Mathematics in Western Culture,* 42.

24 *"a statement used in the premises":* John Daintith and R. David Nelson, *The Penguin Dictionary of Mathematics* (New York: Viking Penguin, 1989), 26.

24 *"that which has no parts":* Kline, *Mathematics in Western Culture,* 42.

24 *"other things being equal":* Boyer, *A History of Mathematics,* 106.

24 *"must be fruitful":* Kline, *Mathematics in Western Culture,* 456.

25 *"derived from premises":* Daintith and Nelson, *The Penguin Dictionary of Mathematics,* 319.

25 *The correctness of the form:* For an excellent description of permissible syllogistic forms, see Rudy Rucker, *Mind Tools* (Boston: Houghton Mifflin, 1987), 200–7.

25 *The abstract quality of this system:* Essentially, in the simplest syllogism there are three properties: The first property mentioned in the conclusion can be termed *S* for subject, the second property mentioned in the conclusion can be termed *P* for predicate, and the third property mentioned in the two premises can be termed *M* for middle term. See generally Rucker, *Mind Tools,* 202.

27 *"Pure mathematics consists entirely"*: Bertrand Russell in "Recent Works on the Principles of Mathematics," *International Monthly* 4 (1901), quoted in *Mathematics,* Devlin, 53 (emphasis added).

28 *"Logic is the art of going wrong"*: Kline, *Mathematics in Western Culture,* 408.

28 *"famous 'mathematical' document"*: Ibid., 328.

28 *"such a common"*: Cohen, *Science and the Founding Fathers,* 123 (emphasis added). One example listed on that page was "A thing cannot be and not be at the same time."

28 *the political equivalents of Euclid's axioms:* There is some speculation that the term *self-evident* was added by Benjamin Franklin. See, for example, Carl L. Becker, *The Declaration of Independence* (New York: Vintage Books, 1958), 142. Even if that were the case, Franklin the scientist would have undoubtedly understood the concept the same way.

29 *"In Disquisitions of every kind"*: Federalist No. 31 (Hamilton).

30 *"[W]e must never forget"*: *McCulloch v. Maryland,* 17 U.S. (4 Wheat.) 316, 407 (1819).

30 *"discern among its 'essential postulates,'"*: *Printz v. U.S.,* 521 U.S. 898 (1997), quoting *Principality of Monaco v. Mississippi,* 292 U.S. 313, 322 (1934).

30 *"partake of the prolixity"*: *McCulloch,* 17 U.S. 407. See also Ronald Dworkin, *Taking Rights Seriously* (Cambridge, Mass.: Harvard University Press, 1977), 179 ("We have volumes full of rules and only a few of principles").

31 *"This great principle"*: *McCulloch,* 17 U.S. 426–27.

31 *"The Court is aware"*: 22 U.S. 1, 221–22 (1824).

32 *"Persecution for the expression of opinions"*: *Abrams v. U.S.,* 250 U.S. 616, 630 (1919) (Justice Oliver Wendell Holmes Jr. dissenting).

34 *Numbers that are not algebraic:* Formally, algebraic numbers are solutions of polynomial equations having rational numbers as coefficients. For example, $x^2 = 1$ has two solutions $x = 1$ and $x = -1$, so both 1 and –1 are algebraic numbers. Similarly, because a solution of $x^2 - 2 = 0$ is $x = \sqrt{2}$, $\sqrt{2}$ is algebraic, even though it is irrational.

34 *it cannot be constructed:* Devlin, *Mathematics: The Science of Patterns,* 121.

34 *"The life of the law"*: Oliver Wendell Holmes Jr., *The Common Law* (1881; reprint, Mineola, N.Y.: Dover, 1991), 5.

34 *"[t]he felt necessities"*: Ibid.

34 *"it plays a role"*: Richard A. Posner, *The Problems of Jurisprudence* (Cambridge: Harvard University Press, 1990), 54–55.

34 *Two infamous Supreme Court cases:* For a powerful analysis of "the role of the American legal process in substantiating, perpetuating, and legitimizing the precept of [racial] inferiority," see A. Leon Higginbotham, *Shades of Freedom* (New York: Oxford University Press, 1996), xxv.

34 Dred Scott: 60 U.S. (19 How.) 393 (1857).

35 "*They had for more than a century*": Ibid., 407 (emphasis added).

35 "*one of the great disasters*": Geoffrey R. Stone, Louis M. Seidman, Cass R. Sunstein, and Mark V. Tushnet, *Constitutional Law* (3d ed.; New York: Aspen Law & Business, 1999), 504. See also Higginbotham, *Shades of Freedom,* 62–67.

35 "*If one element were faulty*": Peterson, *Islands of Truth,* 15.

35 Korematsu: 323 U.S. 214 (1944).

35 "*There were disloyal members*": Ibid., 218 [quoting *Hirabayashi v. U.S.,* 320 U.S. 81, 99 (1943)].

36 "*intentional falsehoods*": See Memorandum of John L. Burling to Assistant Attorney General Herbert Wechsler, September 11, 1944, and Memorandum of Edward J. Ennis, Director of the Alien Enemy Control Unit, Department of Justice to Assistant Attorney General Herbert Wechsler, September 30, 1944, reprinted in *Korematsu v. United States,* 584 F. Supp. 1406 (N.D. Cal. 1984).

36 "*Similar disloyal activities*": *Korematsu,* 323 U.S. 240 (Justice Murphy dissenting).

37 "*The main reasons*": Ibid., 239 (Justice Murphy dissenting). See also Report of the Commission on Wartime Relocation and Internment of Civilians, *Personal Justice Denied* (1982), concluding that the exclusion order resulted from "race prejudice, war hysteria and a failure of political leadership."

37 "*It is better*": Sir William Blackstone, *The Law of England, Book the Fourth* (1807), chap. 27, 358, quoted in *U.S. v. Fatico,* 458 F. Supp. 388, 411 (E.D.N.Y. 1978).

37 *When a mathematical proof is done correctly:* "In human terms, being a proof means having the capacity to completely convince *any* sufficiently educated, intelligent, rational person. . . ." Devlin, *Mathematics,* 38 (emphasis in original).

38 "*empirical judgment that most persons formally accused*": *Bell v. Wolfish,* 441 U.S. 520, 579 (1979) (Justice Stevens dissenting).

38 *pretrial detainees can be confined: Bell,* 441 U.S. 535.

38 *no person can be convicted: In re Winship,* 397 U.S. 358, 364 (1970).

39 "*by a preponderance of the evidence*": See, for example, *U.S. v. One Assortment of 89 Firearms,* 465 U.S. 354, 361-62 (1984).

39 *"probabilistic . . . since the fact finder": Victor v. Nebraska,* 511 U.S. 1, 14 (1994), quoting *In re Winship,* 397 U.S. 370 (Justice Harlan concurring) (emphasis in original).

39 *Type II error:* See generally Ramakant Khazanie, *Elementary Statistics in a World of Applications* (2d ed; Reading, Mass.: Addison-Wesley, 1986), 295-96. In formal statistics, the risk of a Type I error is often indicated by $\alpha$, the risk of a Type II error by $\beta$.

42 *If we convict someone who is innocent: Ballew v. Georgia,* 435 U.S. 223, 234 (1978) (opinion of Justice Blackmun); Paulos, *Innumeracy,* 108; Neil B. Cohen, "Confidence in Probability: Burdens of Persuasion in a World of Imperfect Knowledge," *New York University Law Review* 60 (1985): 385, 410; William S. Laufer, "The Rhetoric of Innocence," *Washington Law Review,* 70 (1995): 329, 394 n. 302. Some have used the opposite labels, describing finding a guilty person innocent as a Type I error and the conviction of an innocent person as a Type II error. See, for example, William M. Landes and Richard A. Posner, "The Economics of Anticipatory Adjudication," *Journal of Legal Studies* 23 (1994): 683, 690. The selection of Type I as a description for wrongful conviction or erroneous acquittal depends on whether one uses an hypothesis of innocence or guilt. For our discussion, it does not matter which characterization is used, as long as it is understood that risks of both Type I and Type II errors are unavoidable and that reducing the risk of one increases the risk of the other.

43 *"a fundamental value determination": In re Winship,* 397 U.S. 372 (Justice Harlan concurring).

43 *it was preferable for ten guilty persons:* Sir William Blackstone, *The Law of England, Book the Fourth,* chap. 27, 358, quoted in *U.S. v. Fatico,* 458 F. Supp. 388, 411.

43 *"Where one party has at stake": Speiser v. Randall,* 357 U.S. 513, 525–26 (1958).

44 *"the social disutility of error": Santosky v. Kramer,* 455 U.S. 745, 790 (1982) (Justice William Rehnquist dissenting).

44 *A defendant was convicted of one count: U.S. v. Watts,* 519 U.S. 148 (1997).

44 *"a middle level": Addington v. Texas,* 441 U.S. 418, 432 (1979). See also *Santosky,* 455 U.S. 756 (terming clear and convincing an "intermediate standard of proof").

44 *The Court has required:* See *Santosky,* 455 U.S. 769; *Addington,* 441 U.S. 424–26; *Woodby v. INS,* 385 U.S. 276, 285 (1966); *Chaunt v. U.S.,* 364 U.S. 350, 353 (1960); *Schneiderman v. U.S.,* 320 U.S. 118, 125 (1943).

44 *"The individual should not be asked": Addington,* 441 U.S. 427; *Santosky,* 455 U.S. 745.

45 *"a natural parent's desire": Santosky,* 455 U.S. 758–59 (internal citations omitted).

45 *"If the Family Court":* Ibid., 790 (Justice Rehnquist dissenting).

45 Cruzan: 497 U.S. 261 (1990).

45 *"protection and preservation of human life":* Ibid., 283.

46 *"An erroneous decision not to terminate":* Ibid., 320 (Justice Brennan dissenting).

46 *While mathematical analysis can tell the Court:* See, for example, Laurence H. Tribe and Michael C. Dorf, *On Reading the Constitution* (Cambridge: Harvard University Press, 1991), 87: "[S]pecifying an appropriate abstraction cannot be accomplished without making value choices."

46 *mathematics is remarkably ineffective:* See generally Laurence H. Tribe, "Trial by Mathematics: Precision and Ritual in the Legal Process," *Harvard Law Review* 84 (1971): 1329, 1374–75. One commentator, noting that "preponderance of evidence" is often described as "more likely than not," explained that "As the standards of proof increase along the continuum from civil to criminal, their descriptions shift from quantitative to qualitative." Paul C. Smith, "The Process of Reasonable Doubt: A Proposed Instruction in Response to *Victor v. Nebraska,*" *Wayne Law Review* 41 (1995): 1811, 1835.

46 *numerical equivalents for differing legal standards:* See, for example, *U.S. v. Shonubi,* 895 F. Supp. 460, 471 (E.D.N.Y. 1995). See also *U.S. v. Fatico,* 458 F. Supp. 388, 403–6, aff'd on other grounds; 603 F. 2d 1053 (2d Cir. 1979), cert. denied, 444 U.S. 1073 (1980). See further Vern Walker, "Direct Inference in the Lost Chance Cases: Factfinding Constraints Under Minimal Fairness to Parties," *Hofstra Law Review* 23 (1994): 247, 257 ("A number of courts have . . . explained the meaning of preponderance . . . using the quantitative terminology of mathematical probability").

46 *Even if one were to assume that the average juror:* A dubious hypothesis for the present. See generally John Allen Paulos, *Innumeracy* (New York: Hill and Wang, 1988).

47 *"No one has yet invented":* J. H. Wigmore, *Evidence* (3d ed; *Boston: Little, Brown,* 1940), vol. 9, sec. 2497, 325. Even my announcement that I am willing to bet that the Chicago Cubs will win the World Series only if someone gives me 50 to 1 odds does not reveal how sure I am that the Cubs will win. The odds at which I am willing to wager repre-

sent a combination of many factors, including both my beliefs and the degree to which I am, at that moment, risk adverse (or risk seeking). The *amount* of my aversion or seeking is not revealed by my willingness to wager. Certainly, a demand for high odds could indicate that I am very risk adverse with a weak belief in the Cubs's invincibility, only slightly risk adverse with a fanatic's confidence in the Cubs, or risk-seeking and wouldn't be at all surprised if the Cubs lost.

47 *"Proof beyond a reasonable doubt is proof":* Federal Judicial Center, Pattern Criminal Jury Instructions (1987), 17–18 (instruction 21), quoted in *Victor,* 511 U.S. 27/Justice Ginsburg concurring).

47 *But great care must be taken:* For a comprehensive analysis of this problem, see Tribe, "Trial by Mathematics," 1329. Among other points, Professor Tribe noted that overreliance on a mathematical-style proof might well "shift the focus away from such elements as volition, knowledge, and intent, and toward such elements as identity and occurrence—for the same reason that the hard variables tend to swamp the soft" (Ibid., 1366). He warned that a jury might have great difficulty balancing hard numbers "against such fuzzy imponderables as the risk of frame-up or of misobservation, if indeed it is not induced to ignore those imponderables altogether" (ibid., 1365).

47 *Mathematical reasoning should not be confused:* "Reducing a complex intelligence or the economy to numbers on a scale, whether I.Q. or GNP, is myopic at best and many times simply ludicrous" (Paulos, *Innumeracy,* 91).

## 2. Majority Rules

48 *"The search of the great minds":* Paul Hoffman, *Archimedes' Revenge* (New York: Norton, 1988), 221.

48 *The American ideal:* Abraham Lincoln, "The Gettysburg Address," in *Lincoln at Gettysburg,* Garry Wills (Thorndike, Me.: Thorndike Press, 1992), 261. For an excellent discussion on how the goal of those framing the Constitution was to promote republicanism as a way to prevent the harms of excessive democracy, see M. N. S. Sellers, *American Republicanism* (New York: NYU Press, 1994), 57–62.

48 *"The fabric of American empire":* Federalist No. 22 (Hamilton).

48 *"Individuals exercise":* Thomas Jefferson, Constitutionality of Residence Bill of 1790, July 15, 1790, in *Founders' Constitution,* Kurland and Lerner, vol. 2, 300.

49 *"[T]he fundamental maxim"*: Federalist No. 22 (Hamilton).

49 *"Logically, in a society"*: *Reynolds v. Sims,* 377 U.S. 533 (1964).

49 *"There is nothing in the language of the Constitution"*: *Gordon v. Lance,* 403 U.S. 1, 6 (1971).

49 *each congressional district must be*: *Wesbury v. Sanders,* 376 U.S. 1 (1964).

50 *"continued frustration of the will"*: *Davis v. Bandemer,* 478 U.S. 109 (1986) (Justice White, plurality opinion).

50 *"Wherever the real power in a Government lies"*: Letter from James Madison to Thomas Jefferson, Oct 17, 1788, quoted in *The Creation of the American Republic: 1776–1787,* Gordon S. Wood (Chapel Hill: University of North Carolina Press, 1969), 410.

51 *"a majority be united"*: Federalist No. 51 (Madison).

51 *"extend the sphere"*: Federalist No. 10 (Madison).

51 *"[Y]ou take in a greater variety of parties"*: Ibid.

51 *"pivotal player"*: Alan D. Taylor, *Mathematics and Politics* (New York: Springer-Verlag, 1995), 69.

52 *"fraction of power"*: The Shapley-Shubik index of a particular voter $p$, denoted SSI($p$), for a set $x$ of voters, is calculated as follows:

$$\text{SSI}(p) = \frac{\text{number of orderings of } x \text{ for which } p \text{ is pivotal}}{\text{number of possible orderings of } x}$$

The Shapley-Shubik index was originally designed to explore the power of a given voter within a committee. See Lloyd Shapley and Martin Shubik, "A Method for Evaluating the Distribution of Power in a Committee System," *American Political Science Review* 48 (1954): 787. My description of this index was derived from Taylor, *Mathematics and Politics,* 69–71.

52 *"concert and execute their plans of oppression"*: Federalist No. 10 (Madison).

52 *"by comprehending in the society"*: Federalist No. 51 (Madison).

54 *at the time of the first census*: The eight smallest states had a combined population of 876,687, which was 24.2 percent of the nation's population of 3,615,920.

54 *As of the 2000 census*: According to the 2000 census, the smallest 26 states had a population of 50,025,674, and the population of the 50 states was 281,424,177. Thus, the 26 smallest states had 17.75 percent of the population of the 50 states.

54 *"If they vote by States"*: Statement of Roger Sherman, July 7, 1787, in

*Debates in the Federal Convention,* 220. See generally Thorton Anderson, *Creating the Constitution* (University Park: Pennsylvania State University Press, 1993), 51–64.

55 *"There are objections":* Statement of James Madison, July 25, 1787, in *Debates in the Federal Convention,* 318–19.

55 *"The people are uninformed":* Statement of Elbridge Gerry, July 19, 1787, in *Debates in the Federal Convention,* 286.

55 *"The right of suffrage":* Statement of James Madison, July 19, 1787, in *Debates in the Federal Convention,* 286.

56 *"A small number of persons":* Ibid. See also Federalist No. 68 (Hamilton), stating that "the immediate election [of the President] should be made by men most capable of analyzing the qualities adapted to the station. . . ."

56 *The first proposal:* Statement of Oliver Elsworth, July 19, 1787, in *Debates in the Federal Convention,* 286–87.

56 *"varied as that it would adjust itself":* Statement of James Madison, July 20, 1787, in *Debates in the Federal Convention,* 288.

56 *It was later proposed:* Statement of Hugh Williamson, July 20, 1787, in *Debates in the Federal Convention,* 289.

57 *A shift of fewer than 10,000 votes:* Similarly, a shift of just 13,000 votes (in Hawaii, Illinois, Missouri, Nevada, and New Mexico) would have resulted in Richard Nixon's defeating John F. Kennedy in 1960.

57 *the winner-take-all feature is* not *required:* See U.S. Const., Art. II, §1, cl. 2. (declaring that the selection of electors shall be "in the manner prescribed by each state legislature . . .").

58 *Physicist Alan Natapoff compared:* See A. Natapoff, "A Mathematical One-Man One-Vote Rationale for Madisonian Presidential Voting Based on Maximum Individual Voting Power," *Public Choice* 88 (1996): 259. See also Will Hivey, "Math against Tyranny," *Discover* 74 (November 1996).

59 *Natapoff proved that districting:* Natapoff's theorem can be stated as follows: Let $u$ represent the measure of the degree to which an election is closely contested, with $u = .5$ indicating a perfectly contested election and $u = 1$ indicating an uncontested election. Also, let $L(X)$ represent the voting power of an individual for a particular system. For every electorate of size $X$ and simple districting scheme $X'$ there is always a critical value $u^*$ such that if $u > u^*$, $L(X') > L(X)$.

60 *chose to campaign in battleground states:* See, for example, Paul West, "Candidates fight for tossup states," *Baltimore Sun*, November 6, 2000, 1A.

60 *if no majority winner emerges:* U.S. Const., Amend. XII.

60 *The few times the House has been involved:* See generally Thomas Galvin, "House Calls and Close Calls," *Congressional Quarterly,* May 23, 1992, 1421.

61 *only a two-person race guarantees:* Of course, there could always be a tie in a two-candidate election. As the size of the electorate increases, though, the likelihood of such a dead heat diminishes.

62 *Consider this scenario involving three candidates:* The following example is derived from Hoffman, *Archimedes' Revenge,* 241–42, discussing Steven Brams and Peter Fishburn, *Approval Voting* (Cambridge, Mass.: Birkhäuser Boston, 1983).

63 *"Condorcet failed to take":* Kasner and Newman, *Mathematics and the Imagination,* 254.

63 *"The obscurity and self-contradiction":* Isaac Toddhunter, *History of the Mathematical Theory of Probability* (1865), 352, quoted in Keith Michael Baker, *Condorcet: From Natural Philosophy to Social Mathematics* (Chicago: University of Chicago Press, 1975), 227.

64 *Condorcet winner:* See, for example, H. P. Young, *"Condoret's Theory of Voting," American Political Science Review* 82 (1986): 123.

65 *It is known as cycling:* See, for example, Saul Levmore, "Parliamentary Law, Majority Decisionmaking, and the Voting Paradox, *Virginia Law Review* 75 (1989): 971, 985.

65 *The Reverend Charles L. Dodgson:* See Duncan Black, *The Theory of Committees and Elections* (Cambridge: Cambridge University Press, 1958), 236. According to Steven Brams, as the number of candidates increases toward infinity, the probability of cycling occurring among three voters increases to 100 [Steven Brams, *Paradoxes in Politics* (New York: Free Press, 1976)]. See also Hoffman, *Archimedes' Revenge,* 230–31.

65 *"I am quite prepared to be told":* C. L. Dodgson, *A Method of Taking Votes on More than Two Issues* (1876; reprinted in Black, *Theory of Committees and Elections*), 224, 230.

65 *Imagine the following preference ranking":* The Borda examples are drawn from Black, *The Theory of Committees and Elections,* 59, 182.

67 *"would be a great advantage":* Ibid. 182.

67 *"The only objection which occurred":* Statement of James Madison, July 25, 1787, in *Debates in the Federal Convention,* 322–23.

67 *"My scheme is only intended":* Ibid. Another problem with the Borda ranking is that it presumes a uniform degree of intensity. When I rank three choices, I may value my top choice far more than twice as much

as my next choice and be practically indifferent between my bottom two choices. Thus, a true valuation of my preference may be lacking (see Black, *The Theory of Committees and Elections,* 65).

68 *that every possible voting scheme:* Kenneth J. Arrow, *Social Choice and Individual Values* (New York: Wiley, 1951), 42–59. See generally Maxwell L. Stearns, "The Misguided Renaissance of Social Choice," *Yale Law Journal* 103 (1994): 1219.

68 *"No voting rule":* Jerry Mashaw, *Greed, Chaos, and Governance* (New Haven: Yale University Press, 1997), 12–13.

69 *we may find several plausible answers:* Black, *The Theory of Committees and Elections,* 57.

69 *Consider the following situation with four parties:* The numbers used in this example are derived from Paulos, *Beyond Numeracy,* 264.

70 *"How we should be democratic":* Ibid., 265. See also Mashaw, *Greed, Chaos, and Governance,* 15 (stating, "Voting theory teaches that majoritarian democracy is a necessarily compromised process and that any institutional design chosen to promote democracy will have a problematic relationship to our normative aspirations").

## 3. The Positive Value of Consensus

71 *"[T]he right of expulsion":* James Madison, August 10, 1787, in *Debates in the Federal Convention,* 378.

71 *"bare majority":* See, for example, Federalist No. 73 (Hamilton) (emphasis added): "[I]t will not often happen that improper views will govern so large a proportion as two thirds of both branches of the legislature at the same time. . . . It is at any rate far less probable that this should be the case, than that such views should taint the resolutions and conduct of *a bare majority.*"

72 *"Except this original contract":* Jean-Jacques Rousseau, *Social Contract,* b. 4, chap. 2 (1762). See generally John Heinberg, "Theories of Majority Rule," *American Journal of Political Science Review* 26 (1932): 452.

72 *We are not to suppose": Wood v. Lucy, Lady Duff-Gordon,* 118 N.E. 214, 214 (N.Y. 1917).

72 *"to sacrifice to its ruling passion":* Federalist No. 10 (Madison).

72 *"The very purpose of a Bill of Rights": West Virginia State Board of Education v. Barnette,* 319 U.S. 624, 638 (1943).

73 *there were five different provisions:* Two subsequent amendments have also required a two-thirds majority: Two-thirds of both Houses were

necessary for removing the legal disabilities of those elected officials who supported the South during the Civil War, Amendment XIV, §3, and two-thirds of both Houses are needed to determine presidential incapacity, Amendment XXV, §4.

73 *to override a presidential veto:* U.S. Const., Art. I, §7, cl. 3 (override of veto); U.S. Const. Art. V (proposing amendments). An even greater fraction of states, three-fourths, is required to ratify an amendment.

73 *expelling a member:* U.S. Const., Art I, §5, cl. 2.

73 *conviction after impeachment:* U.S. Const., Art. I, §3, cl. 6 (conviction after impeachment); U.S. Const., Art. II, §2, cl. 2 (treaties). For the House of Representatives to impeach, only a majority vote is needed, U.S. Const., Art. I, §2, cl. 5.

73 *three-fourths of the states:* U.S. Const., Art. V.

73 *impeachment by a majority:* The impeachment provisions are contained in U.S. Const., Art. I, §2. cl. 5 (House of Representatives power to impeach); Art. I, §3, cl. 6 (Senate power to try all impeachments, requiring two-thirds concurrence); and Art. II, §4 (president, vice president, and all civil officers subject to impeachment). The proposal to permit conviction by a majority of the Senate is described in *Debates in the Federal Convention,* September 4, 1787, 508. The vote in favor of Gouverneur Morris's proposal to change the requirement for conviction from majority to two-thirds was 9 states to 2 (ibid., September 8, 1781, 537).

74 *"[N]o offender can escape the danger":* Archibald Maclaine, North Carolina Ratifying Convention, July 25, 1788, in Kurland and Lerner, *Founders' Constitution,* vol. 2, 163.

74 *"If a mere majority":* Story, *Commentaries on the Constitution,* 277-78.

74 *the primary ground for the impeachment proceeding:* See generally Laurence H. Tribe, *American Constitutional Law* (2d ed; Eagan, Minn.; Foundation Press, 1988), 291–92; M. Benedict, *The Impeachment and Trial of Andrew Johnson* (New York: Norton, 1973). The Tenure of Office Act of 1867 was held to be unconstitutional in *Myers v. U.S.,* 272 U.S. 52 (1926).

75 *"[I]f the guilt of a public officer":* Story, *Commentaries on the Constitution,* 277–78.

75 *"where a bare majority has carried":* James Iredell, North Carolina Ratifying Convention, July 26, 1788, in Kurland and Lerner, *Founders' Constitution,* vol. 2, 402. This view was echoed by Hamilton: "[The veto] furnishes an additional security against the enaction of improper laws. It establishes a salutary check upon the legislative body, calcu-

lated to guard the community against the effects of faction, precipitancy, or of any impulse unfriendly to the public good, which may happen to influence a majority of that body" (Federalist No. 73).

75 *"defend the Executive Rights":* James Madison, September 12, 1787, in *Debates in the Federal Convention,* 556.

75 *"an absolute negative":* James Wilson, June 4, 1787, in *Debates in the Federal Convention,* 51.

76 *"No one man could be found":* Roger Sherman and George Mason, June 4, 1787, in *Debates in the Federal Convention,* 52, 54.

76 *"[I]f a proper proportion":* James Madison, June 4, 1787, in *Debas in the Federal Convention,* 53.

76 *"The excess rather than the deficiency":* Gouveneur Morris, September 12, 1787, in *Debates in the Federal Convention,* 555.

76 *"puts too much in the power":* Hugh Williamson, September 12, 1787, in *Debates in the Federal Convention,* 554.

76 *"a few Senators having hopes":* Elbridge Gerry, September 12, 1787, in *Debates in the Federal Convention,* 555. As Joseph Story, would later write, "The larger the number required to overrule the executive negative, the more easy it would be for him to exert a silent and secret influence to detach the requisite number in order to carry his object. . . . [I]t would be easy to defeat the most salutary measures, if a combination of a few states could produce such a result." Story, *Commentaries on the Constitution,* 323–24.

76 *"little arithmetic was necessary":* George Mason, September 12, 1787, in *Debates in the Federal Convention,* 555.

76 *"we must":* James Madison, September 12, 1787, in *Debates in the Federal Convention,* 556.

76 *"on the rule":* James Madison, September 12, 1787, in *Debates in the Federal Convention,* 556.

76 *[A]n expression of opinion by two thirds":* Story, *Commentaries on the Constitution,* 324.

77 *a popular bill to regulate cable television rates:* Cable Television Consumer Protection Act of 1992, U.S. Code, vol. 47, sec. 521–59. See Adam Clymer, "Congress Rebuffs Bush in Override of Cable TV Veto," *New York Times,* October 6, 1992, A1.

77 *President Clinton's 1996 veto:* "Veto Is Sustained On Abortion Ban," *New York Times,* September 27, 1996, A1.

77 *"'Tis our true policy":* George Washington, "Farewell Address," September 19, 1796, in Kurland and Lerner, *Founders' Constitution,* vol. 1, 681, 685.

77 *"He must either have been very unfortunate":* Federalist No. 64 (Jay).

78 *"If the majority cannot be trusted":* James Wilson, September 8, 1787, in *Debates in the Federal Convention,* 532.

78 *"the power of a minority":* James Wilson, September 7, 1787, in *Debates in the Federal Convention,* 530.

78 *Nonetheless, his proposal that treaties:* September 8, 1787, in *Debates in the Federal Convention,* 534.

78 *"This will be less security than 2/3":* Hugh Williamson, September 8, 1787, in *Debates in the Federal Convention,* 534.

78 *"The security essentially intended by the Constitution":* Federalist No. 66 (Hamilton) (emphasis added).

78 *ten treaties have received a majority:* See, for example, George H. Haynes, *The Senate of the United States* (1938; reprint, New York: Russell and Russell, 1960), 659.

78 *The most prominent of these:* Actually, the vote on ratification of the Treaty of Versailles without any conditions failed by a vote of 38 yeas to 53 nays (Congressional Record, November 19, 1919, 58: 8803). It was a later vote on the treaty with numerous reservations that garnered a majority, 49 to 35, but fell seven votes short of the two-thirds majority (Congressional Record, March 19, 1920, 59: 4599). For the story of Senate treatment of the Treaty of Versailles, see Haynes, *The Senate of the United States,* 694–703.

79 *"grave-yard of treaties":* Louis Henkin, *Foreign Affairs and the U.S. Constitution* (2d ed.; Oxford: Oxford University Press, 1996), 178.

79 *The last time a major treaty was rejected:* Congressional Record, 1935, 79: 1146–47. See generally Haynes, *The Senate of the United States,* 707–13.

79 *"[T]he rule's most calamitous effects":* Haynes, *The Senate of the United States,* 660.

79 *"[I]t heartens any tiny group":* Ibid.

79 *the two major forms of executive agreements:* See generally Henkin, *Foreign Affairs and the U.S. Constitution,* 215–29.

80 *Between 1960 and 1983:* In this context, *executive agreements* includes those which were entered into by the president alone, as well as those which went through both houses by majority vote. See *Treaties and Other International Agreements: The Role of the United States Senate,* S. Rep. No. 205, 98th Cong., 2d sess., 1984.

80 *"the very delicate, plenary and exclusive power": U.S. v. Curtiss-Wright Export Corp.,* 299 U.S. 304, 319–20 (1936). See also *U.S. v. Belmont,* 301 U.S. 324, 330–31 (1937) ("the Executive has the authority to speak as the sole organ of [the national] government").

80 *immediate unilateral presidential action:* See, for example, *Dames and Moore v. Regan,* 453 U.S. 654 (1981) (upholding the sole executive agreement to settle claims with Iran as part of resolution of the Iranian hostage crisis).

80 *"Constitutional doctrine to justify Congressional-Executive agreements":* Henkin, *Foreign Affairs and the U.S. Constitution,* 216.

80 *"Neither Congress, nor Presidents":* Ibid., 217. See also Bruce Ackerman and David Golove, "Is NAFTA Constitutional?" Harvard Law Review 108 (1995): 801, 873–96.

80 *Because this agreement need only be approved:* Henkin, *Foreign Affairs and the U.S. Constitution,* 217.

80 *"On certain occasions":* Wallace Mitchell McClure, International Executive agreements. *Democratic Procedure under the Constitution of the United States* (New York: AMS Press, 1941), 4.

81 *"unaskable":* Ackerman and Golove, "Is NAFTA Constitutional?" 802. See also Myres S. McDougal and Asher Lans, "Treaties and Congressional-Executive or Presidential Agreements: Interchangeable Instruments of National Policy" (pts. 1 and 2), *Yale Law Journal* 54 (1945): 181, 524.

81 *their constitutionality has never:* The Court has addressed related issues, however, without implying that it was particularly troubled by congressional-executive agreements [see, example, *B. Altman & Co. v. U.S.,* 224 U.S. 583 (1912) (construing a congressional-executive agreement as a treaty for statutory purposes); *Cotzhausen v. Nazro,* 107 U.S. 215 (1882) (postal convention held binding though made by congressional-executive agreement)]. Other cases have involved sole executive agreements [see *Dames and Moore,* 453 U.S. 654 (upholding the constitutionality of a sole executive agreement); *Weinberger v. Rossi,* 456 U.S. 25 (1982) (construing a sole executive agreement as a treaty for statutory purposes); *Belmont,* 301 U.S. 324 (upholding presidential recognition of the Soviet Union)].

81 *not subject to judicial review:* See, for example, Henkin, *Foreign Affairs and the U.S. Constitution,* 457 n. 45 ("It may be urged that in view of the possible international consequences of declaring a treaty provision unconstitutional, courts should not consider the validity of a treaty provision but treat that as a political question not for judicial determination"). As Justice Chase wrote in 1796, "If the court possesses a power to declare treaties void, I shall never exercise it, but in a very clear case indeed" [*Ware v. Hylton,* 3 U.S. 199, 237 (1796)].

81 *"constitutional purism":* Thomas M. Franck and Edward Weisband,

*Foreign Policy by Congress* (New York: Oxford University Press, 1979), 149.

81 *"[I]n requiring that all treaties":* J. W. Foster, *The Practice of Diplomacy* (New York: Houghton Mifflin, 1906), 263.

81 *"the advantage of numbers":* Federalist No. 75 (Hamilton). For an important discussion of possible interpretation of the Treaty Clause which preserves the "advantage of numbers," see Tribe, "Taking Text and Structure Seriously," 1264–69.

## 4. The First Veto

82 *"No invasions of the constitution":* Thomas Jefferson, "Opinion on the Bill Apportioning Representation," in *The Papers of Thomas Jefferson,* vol. 23, 370, 373.

82 *"shall be apportioned":* U.S. Const., Amend. XIV; see U.S. Const., Art. I, §2. The Fourteenth Amendment eliminated the requirement that the number be determined by adding the number of free citizens to three-fifths the number of slaves, but retained the language that apportionment be "according to their respective numbers."

82 *let's look at the ideal situation:* Much of the mathematical analysis in this section is derived from Hoffman, *Archimedes' Revenge,* 250–60.

83 *"fair share":* Michel L. Balinski and H. Peyton Young, *Fair Representation* (1982), 14.

84 *"The problem is that although the loyalties":* Hoffman, *Archimedes' Revenge,* 250.

84 *first census:* Balinski and Young, *Fair Representation,* 11.

85 *the Constitution requires that each state:* U.S. Const., Art. I, §2.

85 *Congress approved a bill:* Balinski and Young, *Fair Representation,* 15.

86 *"any principle at all":* Thomas Jefferson, "Opinion on the Bill Apportioning Representation."

86 *"according to any . . . crotchet":* Ibid., 375.

86 *If the total so allocated was less:* A simplified model shows how this worked. Picture a 26-seat House, with a total population of 26,000 divided among 5 states. The quota for each state is determined by multiplying its population by the ratio of

$$\frac{\text{Total seats}}{\text{Nat'l population}} = \frac{26}{26{,}000} = \frac{1}{1{,}000}$$

In the first round of allocation, each state receives as its number of representatives the amount equal to the whole number portion of its quota. In the final round, the extra representative is allocated to the state with the highest fractional remainder.

| State | Population | Ideal Number of Representatives | First Round | Final Round |
|---|---|---|---|---|
| A | 9,061 | 9.061 | 9 | 9 |
| B | 7,179 | 7.179 | 7 | 7 |
| C | 5,259 | 5.259 | 5 | 5 |
| D | 3,319 | 3.319 | 3 | 4 |
| E | 1,182 | 1.182 | 1 | 1 |
| Total | 26,000 | 26 | 25 | 26 |

*Table from Hoffman,* Archimedes' Revenge, *251.*

86 *This number was calculated:* Jefferson's plan for the above 26-seat House example, would have utilized the number 906.1 to yield the proper results.

| State | Population | After Division by Divisor of 906.1 | Allotment by Ignoring Remainders |
|---|---|---|---|
| A | 9,061 | 10.000 | 10 |
| B | 7,179 | 7.923 | 7 |
| C | 5,259 | 5.804 | 5 |
| D | 3,319 | 3.663 | 3 |
| E | 1,182 | 1.304 | 1 |
| Total | 26,000 | | 26 |

*Table from Hoffman,* Archimedes' Revenge, *251.*

86 *A direct comparison:* Balinski and Young, *Fair Representation,* 15, 19.

87 *An examination of census data:* Ibid., 74.

88 *"being a shrewd man of science":* Hoffman, *Archimedes' Revenge,* 254.

88 *President Washington issued the first veto:* President Washington's veto

message focused on Jefferson's analysis that the bill used two divisors and his calculation that Connecticut's allotment of 8 representatives exceeded the constitutional limit of 1 for every 30,000 persons. The full veto message states:

*I have maturely considered the act passed by the two Houses entitled "An act for an Apportionment of Representatives among the several States, according to the first Enumeration," and I return it to your House, wherein it originated, with the following objections:*

*First. The Constitution has prescribed that Representatives shall be apportioned among the several States according to their respective numbers; and there is no one proportion or divisor which, applied to the respective numbers of the States, will yield the number and allotment of Representatives proposed by the bill.*

*Second. The Constitution has also provided that the number of Representatives shall not exceed one for every thirty thousand; which restriction is, by the context, and by fair and obvious construction, to be applied to the separate and respective numbers of the States; and the bill has allotted to eight of the States more than one for every thirty thousand.*

—*3* Annals of Congress, *1792, 539.*

88 U.S. Department of Commerce: 503 U.S. 442, 444 (1992). This was actually the fifth method used to apportion representatives. The Jefferson plan was used from 1792 to 1830. It was replaced in 1840 by a plan proposed by Daniel Webster (which unlike the Jefferson plan did not ignore remainders but rounded fractional remainders above 1/2 up to the next whole number). The Webster plan was also used in 1910 and 1930. The Hamilton plan was used from 1850 to 1900, and the method of equal proportions has been used since 1940 (ibid., 450–52) Congress failed to conduct a new apportionment in 1920 [see Zechariah Chafee, "Congressional Reapportionment," *Harvard Law Review* 42 (1929): 1015, 1015–17].

88 *A different method:* The method of the harmonic mean works like the method of equal proportions but uses a different series of divisors. For the method of the harmonic mean, the divisors follow the pattern $2n\ (n - 1)$; thus the priority list is made by dividing state populations first by 4 (which is $2 \times 2 \times 1$), then 12 ($2 \times 3 \times 2$), then 24 ($2 \times 4 \times 3$), and so on. See *U.S. Department of Commerce v. Montana,* 503 U.S. 455 n. 26.

88 *Montana's gain:* Attempts to explain this method can be found ibid.; Chafee, "Congressional Reapportionment," 1029 n. 39.

88 *With the method of the harmonic mean: U.S. Department of Commerce,* 503 U.S. 462 n. 40. The deviations from each of the different methods were as follows:

**Percentage Difference of Each District from Ideal**

| *State* | *Equal Proportions* | *Harmonic Means* |
|---|---|---|
| Montana | 40.4% | 42.5% |
| Washington | 5.4% | 6.7% |

**Total Amount Difference from Ideal**

| *State* | *Equal Proportions* | *Harmonic Means* |
|---|---|---|
| Montana | 231,189 | 170,638 |
| Washington | 29,361 | 38,527 |
| Total Difference | 260,550 | 209,165 |

88 *"neither mathematical analysis nor constitutional interpretation":* Ibid., 463.

89 *"good faith choice":* Ibid., 464.

89 *"maintain partisan political advantage":* Ibid., 464 n. 42.

89 *"One Democrat is quoted": Congressional Record,* 1941, 87: 1126 (statement of Representative Clare E. Hoffman, February 18, 1941).

## 5. What Does Equality Equal?

91 *"A classification having some reasonable basis": Lindsley v. Natural Carbonic Gas Co.,* 220 U.S. 61, 78 (1911).

91 Carolene Products Co.: 304 U.S. 144, 152, 152 n. 4 (1938).

92 *Professor Bruce Ackerman has pointed out:* Bruce A. Ackerman, "Beyond Carolene Products," *Harvard Law Review* 98 (1985): 713.

92 *A look at any well-heeled:* See also Daniel A. Farber and Philip P. Frickey, "Is Carolene Products Dead?" *California Law Review* 79 (1991): 686.

93 *Judicial scrutiny of democratically enacted laws:* Professor Louis Lusky, Justice Stone's law clerk during the drafting of *Carolene Products,* has described his view of *discrete and insular* as applying "to groups that are

not embraced within the bond of community kinship but are held at arm's length by the group or groups that possess dominant political and economic power" [Louis Lusky, "Footnote Redux: A Carolene Products Reminiscence," *Columbia Law Review* 82 (1982): 1093, 1105 n. 72].

93 Craig v. Boren: 429 U.S. 190 (1976).

93 *"Certainly if maleness"*: Ibid., 201–2.

93 *[I]t does not seem to me"*: Ibid., 214 (Justice Stevens concurring).

94 *Wasn't an insult:* See also *South Dakota v. Dole,* 483 U.S. 203 (1987) (upholding the power of Congress to condition some highway funds on states' raising the drinking age to 21).

94 *"Judgments about people": J. E. B. v. Alabama,* 511 U.S. 127, 140 (1994).

94 *"may not be used, as they once were": U.S. v. Virginia,* 518 U.S. 515, 534 (1996).

94 *"To say that gender makes no difference": J. E. B.,* 511 U.S. 149 (Justice O'Connor concurring).

94 *Statistics cannot justify:* See *Stanton v. Stanton,* 421 U.S. 7, 14–15 (1975), which struck down a Utah statute requiring parents to support male children until they were 21 but females only until age 18. The Court rejected Utah's claim that men, more than women, need a good education so they could provide for their families: "[N]o longer is the female destined solely for the home and the rearing of the family, and only the male for the marketplace and the world of ideas. Women's activities and responsibilities are increasing and expanding."

94 *even statistically sound sex discrimination:* As the Court has stated, "[P]roving broad sociological propositions by statistics is a dubious business, and one that inevitably is in tension with the normative philosophy that underlies the Equal Protection Clause." *Craig,* 429 U.S. 204.

94 *Using statistics to prove unconstitutional discrimination:* Some important articles on this topic include Julie Lamber et al., "The Relevance of Statistics to Prove Discrimination: A Typology," *Hastings Law Journal* 34 (1983): 553; Michael Finkelstein, "The Application of Statistical Decision Theory to the Jury Discrimination Cases," *Harvard Law Review* 80 (1966): 338; Peter Sperlich and Martin Jaspovice, "Methods for the Analysis of Jury Panel Selections: Testing for Discrimination in a Series of Panels," *Hastings Constitutional Law Quarterly* 6 (1979): 787.

94 Yick Wo: 118 U.S. 356 (1886).

95 *"with a mind so unequal and oppressive"*: Ibid., 373.

95 Gomillion: 364 U.S. 339, 340 (1960).

95 *"essay in geometry and geography":* Ibid., 341.

95 *"Disproportionate impact is not irrelevant": Washington v. Davis,* 426 U.S. 229 (1976).

95 *More than numbers are necessary:* The Court looks toward factors such as historical background, sequence of events, departures from normal procedural sequences, and contemporaneous statements, along with the statistical disparity, to determine whether there was a discriminatory intent [*Village of Arlington Heights v. Metropolitan Housing Development Corp.,* 429 U.S. 252 (1977)].

95 McClesky: 481 U.S. 279 (1987).

95 *"was assessed in 22% of cases":* Ibid., 286. The Court did not say that it agreed with the mathematical validity of the study. Instead, the Court stated that even if the study were accurate, it would still not prove a constitutional violation.

96 *"consonant with our understanding":* Ibid., 328 (Justice Brennan dissenting).

96 *"[W]e decline to assume":* Ibid., 313.

96 *The ultimate, bottom-line value judgment questions:* Compare, for example, Van Alstyne, "Rites of Passage: Race, The Supreme Court, and the Constitution," *University of Chicago Law Review* 46 (1979): 775, 809 (stating that "one gets beyond racism by getting beyond it now; by a complete, resolute, and credible commitment *never* to tolerate in one's own life—or in the life or practices of one's government—the differential treatment of other human beings by race") with T. Alexander Aleinkoff, "A Case for Race-Consciousness," *Columbia Law Review* 91 (1991): 1061, 1110 (stating that recognizing race "validates the lives and experiences of those who have been burdened because of their race. White racism has made 'blackness' a relevant category in our society. Yet colorblindness seeks . . . to tell blacks that they are no different from whites, even though blacks as blacks are persistently made to feel that difference").

98 *this distinction equates to the question:* It is important to keep in mind that these pictures ignore the fact that the population of a given race is not monolithic. Their simplified nature also obscures the multiplicity of races and ethnic groups in the United States. See, for example, Pete Wilson, "The Majority-Minority Society," in *The Affirmative Action Debate,* ed. George E. Curry (Reading, Mass.: Addison-Wesley, 1996), 167–74, 277, 283; Harry P. Pachon, "Invisible Latinos: Excluded from Discussions of Inclusion," ibid., 184–90.

98 *school systems and employers continued to be found guilty:* See, for exam-

ple, *Swann v. Charlotte-Mecklenburg Board of Education,* 402 U.S. 1 (1971) ("Deliberate resistance of some to the Court's mandate has impeded the good faith efforts of others to bring school systems into compliance"); *U.S. v. Paradise,* 480 U.S. 92, 162–63 (1987) (upholding the remedial order of the district court that found in 1983 that the effects of the police departments deliberate racial discrimination "remain pervasive and conspicuous at all ranks above the entry level"); *Local 28, Sheet Metal Workers International Association v. EEOC,* 478 U.S. 421 (1986) (noting the court finding in 1975 that the union deliberately discriminated against nonwhites).

98 *One also need not be a Pollyanna:* See, for example, A. Barry Rand, "Diversity in Corporate America," in *The Affirmative Action Debate,* Curry, 65, 68 (noting that African American employment at Xerox has increased from under 3 percent in 1964, to almost 8 percent in 1974, to 14 percent in 1996).

99 *a fundamental point:* By focusing on the question of quality, I recognize that I am not dealing with several other important questions including the benefits of diversity and the harm caused by any use of explicit racial preferences [see, for example, *University of California v. Bakke,* 438 U.S. 265, 312 (1978) (opinion of Justice Powell) (stating that "The atmosphere of speculation, experiment and creation . . . is widely believed to be promoted by a diverse student body . . ."); *City of Richmond v. J. A. Croson Co.,* 488 U.S. 469, 495 (1989) (Justice O'Connor, plurality opinion) (stating that such use "assures that race will always be relevant in American [life]")]. That is not to say these are not vitally important questions. An additional issue, which in a sense runs parallel (in a figurative sense) to the affirmative action debate, is the question of improving the general quality of applicants. This is universally seen as a virtue, hopefully a virtue worth pursuing aggressively, but it does not directly confront the concern of all sides in the debate that the selection model accurately assess all applicant's qualifications.

100 *proxy:* See, for example, *Craig,* 429 U.S. 204.

100 *Both a good model:* Hoffman, *Archimedes' Revenge,* 157 ("That is the beauty of a mathematical representation: it need only preserve what is relevant to the situation at hand. Free of distracting irrelevancies, the mathematician is better able to concentrate on the problem").

100 *"[T]he certainty of mathematical conclusions":* Paulos, *Beyond Numeracy,* 149.

100 *we will borrow a form of analysis:* For a discussion of Condorcet's idea of head-to-head comparisons, see Chapter 2.

102 *we must divide the above proportions:* Note that opponents of affirmative action say that this step represents one of the very flaws of affirmative action, forcing and maintaining race consciousness. Proponents respond that race consciousness is endemic in society today; it would exist regardless of the use of affirmative action.

103 *quotas have long been declared unconstitutional:* See *Bakke,* 438 U.S. 289 (opinion of Justice Powell); *Croson Co.,* 488 U.S. 499.

103 "*[R]ace or ethnic background may be deemed a 'plus'*": *Bakke,* 438 U.S. 317 (opinion of Justice Powell).

103 "*candidates who are* less qualified": Charles T. Canady, "The Meaning of American Equality," in *The Affirmative Action Debate,* Curry, 277, 283 (emphasis in original).

104 "*[A]s the basis for imposing discriminatory* legal *remedies*": *Wygant v. Jackson Board of Education,* 476 U.S. 267, 276 (1986) (Justice Powell, plurality opinion) (emphasis in original).

104 "*You have but to ask any African-American*": Deval L. Patrick, "Standing in the Right Place," in *The Affirmative Action Debate,* Curry, 137, 138.

105 "*'completely unrealistic' assumption*": *Croson,* 488 U.S. 507.

105 "*chronic and apparently immutable handicaps*": *Adarand Constructors,* 515 U.S. 241 (Justice Thomas concurring).

106 "*the son of an African-American schoolteacher*": Victoria Valle, "Sitting In for Diversity," in *The Affirmative Action Debate,* Curry, 210, 214.

## 6. Game Theory and the Constitution

109 "*One more question*": This exchange is from an interview with James Dean conducted in 1955, shortly before his death in an automobile accident. I discovered this excerpt online at www.moviesounds.com/rebel.html.

109 *Game theory is a formal mathematical field:* See, for example, Douglas G. Baird, Robert H. Gertner, and Randal C. Piker, *Game Theory and the Law* (Cambridge: Harvard University Press, 1994), 308.

109 "*conflict between thoughtful and potentially deceitful opponents*": William Poundstone, *Prisoner's Dilemma* (New York: Anchor Books, 1992), 6.

109 *it presents ways to think:* For an excellent introduction to the topic, see Stephen W. Salant and Theodore S. Sims, "Law and Economics: Game Theory and the Law: Ready for Primetime?" *Michigan Law Review* 94 (1996): 1839.

109 "*a complete description of a particular way*": Poundstone, *Prisoner's Dilemma,* 48.

110 *"their decisions in an essentially amoral":* John L. Casti, *Five Golden Rules* (New York: Wiley, 1996), 41. Oliver Wendell Holmes advised a similar construct for viewing the law: "We look at the law as it would be regarded by one who had no scruples against anything which he could do without incurring legal consequences" (*The Common Law,* 311).

110 *Game theory can provide valuable insight:* Richard W. Hamming, quoted in Casti, *Five Golden Rules,* 41.

111 *a strategy is optimal:* This is sometimes known as the Minimax Theorem, because as one player tries to maximize his winnings, the other tries to keep that maximum to a minimum. For some games, the optimal strategy is a mixed strategy, meaning that the moves are selected at random, on the basis of a predetermined probability.

112 Marbury: 5 U.S. (1 Cranch) 137 (1803).

114 *"It was obvious to all":* Julius J. Marke, *Vignettes of Legal History* (South Hackensack, N.J.: Fred B. Rothman & Company, 1965), 12.

115 *"men are ambitious":* Federalist No. 6 (Hamilton).

115 *"the principle of the unitary national market": West Lynn Creamery, Inc. v. Healy,* 512 U.S. 186, 193 (1994).

115 *"Each State, or separate confederacy":* Federalist No. 7 (Hamilton).

115 *"implied Dormant Commerce Clause:* The term *Dormant Commerce Clause* evolved from language in *Willson v. Black Bird Creek Marsh Co.,* 27 U.S. 245, 252 (1829), discussing whether a particular state law was "repugnant to Congress's power to regulate commerce in its dormant state." See generally Jacques Leboeuf, "The Economics of Federalism and the Proper Scope of the Federal Commerce Power," *San Diego Law Review* 31 (1994): 555, 610 n. 252.

116 *"whose purpose or effect": South Carolina Highway Department v. Barnwell Bros.,* 303 U.S. 177, 185 (1938).

116 *Some modern commentators:* See, for example, Louise Weinberg, "Against Comity," *Georgetown Law Journal* 80 (1991): 53, 55; Jenna Bednar and William N. Eskridge Jr., "Steadying the Court's Unsteady Path: A Theory of Judicial Enforcement of Federalism," *Southern California Law Review* 68 (1995): 1447, 1471; Leboeuf, "The Economics of Federalism and the Proper Scope of the Federal Commerce Power," 574; Peter Yu, "Note: To Form a More Perfect Union?: Federalism and Informal Interstate Cooperation," *Harvard Law Review* 102 (1989): 842, 845.

Albert Tucker is credited with coining the term *prisoner's dilemma,* though the basic dilemma was first described by Merril Flood of the RAND Corporation in 1951 (See Casti, *Five Golden Rules,* 34).

117 *In repeated versions of the game:* In fact, in a prisoner's dilemma tournament for different strategies, the winner was a strategy called tit for tat. It started off cooperating, and then responded to what the opposing player had done on the previous round: if the other player had cooperated, it cooperated; if the other had defected, it would defect. This simple strategy defeated all comers, including many of far greater sophistication (see Poundstone, *Prisoner's Dilemma,* 239).

118 *when New Jersey prevented residents of Pennsylvania: City of Philadelphia v. New Jersey,* 437 U.S. 617 (1978).

119 *a trade war:* See, for example, *A & P Tea Co. v. Cottrell,* 424 U.S. 366 (1976) (holding that retaliation is impermissible, even if motivated by a desire to end trade barriers).

119 *federal intervention:* See, for example, Larry Alexander and Frederick Schauer, "On Extrajudicial Constitutional Interpretation," *Harvard Law Review* 110 (1997): 1359, 1371 (stating that law can create the "benefits of inducing socially beneficial cooperative behavior and providing solutions to Prisoner's Dilemmas and other problems of coordination").

119 *the power "to regulate Commerce":* U.S. Const., Art. I, §8, cl. 3.

119 *"[T]hese restraints are individually too petty": Duckworth v. Arkansas,* 314 U.S. 390, 400 (1941) (Justice Jackson concurring).

119 *The clearest example is a protective tariff:* See, for example, *West Lynn Creamery, Inc.,* 512 U.S. 186 (1994).

119 *"basically a protectionist measure": City of Philadelphia,* 437 U.S. 624.

120 *"burden falls principally upon those without": Barnwell Bros.,* 303 U.S. 185.

120 *Imagine that Minnesota:* This example is derived from *Minnesota v. Clover Leaf Creamery Co.,* 449 U.S. 456 (1981).

120 *"the existence of major in-state interests": Clover Leaf Creamery Co.,* 449 U.S. 473 n. 17. In this case, the Court upheld the law, noting that there were in-staters, as well as out-of-staters, who were adversely affected by the law.

121 West Lynn Creamery, Inc.: 512 U.S. 186 (1994).

121 *"[O]ne would ordinarily expect":* Ibid., 200–1.

121 *A Maryland law banning companies: Exxon Corp. v. Maryland,* 437 U.S. 117 (1978). The in-state and out-of-state figures are from Justice Blackmun's dissent, ibid., 138 (Justice Blackmun dissenting).

121 *"The fact that the burden":* Ibid., 126.

122 *"Where the statute regulates even-handedly": Pike v. Bruce Church, Inc.,* 397 U.S. 137, 142 (1980).

122 *Legislators need not be rational:* See, for example, *U.S. Railroad Retirement Board v. Fritz,* 449 U.S. 166, 179 (1980) (stating it was irrelevant if Congress "was unaware of what it had accomplished . . .").

122 *the cost of Iowa's ban on long trucks: Kassel v. Consolidated Freightways Corp.,* 450 U.S. 662 (1981) (striking down Iowa's ban on 65-foot-long trucks).

122 *"I do not know what qualifies us": CTS Corp. v. Dynamics Corp. of America,* 481 U.S. 69, 95 (Justice Scalia concurring).

123 *if a law directly discriminates:* See, for example, *Hughes v. Oklahoma,* 441 U.S. 322, 337 (1979).

123 *"[s]hielding in-state industries": Maine v. Taylor,* 477 U.S. 131, 148 (1986).

123 *"would benefit only a few Iowa-based companies": Kassel,* 450 U.S. 666 (plurality opinion).

123 *"asserted justification . . . is merely a pretext": Kasel,* 450 U.S. 692 (Justice Rehnquist dissenting). See also *Maine v. Taylor,* 477 U.S. 149 (upholding the ban on imported baitfish because "there is little reason in this case to believe that the legitimate justifications the State has put forward are merely a sham or a post hoc rationalization").

123 *"remarkably little to further": Hunt v. Washington State Apple Advertising Commission,* 432 U.S. 333, 353 (1977).

124 *"illusory, insubstantial, or nonexistent": Kassel,* 450 U.S. 681 n. 1 (Justice Brennan dissenting).

## 7. Multidimensional Thinking

125 *"[T]he rational person who has grasped":* Kline, *Mathematics in Western Culture,* 430.

125 *"One theme that recurs time and again":* Osserman, *Poetry of the Universe,* 61.

126 *It took many centuries:* Ibid., 61–62.

126 *The infallibility of Euclid:* In the words of Morris Kline, "Century after century buttressed logic with experience and common sense with tradition, until Euclid's system acquired inviolate sanctity. By 1800 educated people were far more likely to swear by the theorems of Euclid than by any statement in the Bible" (*Mathematics in Western Culture,* 410). See also Osserman, *Poetry of the Universe,* 63 (saying, "By the nineteenth century, Euclid's geometry was two thousand years old, and had been a central component of a general education for centuries; it was also the prototype of clear thinking and logical reasoning").

126 *The first four of Euclid's five postulates:* See, for example, Philip J. Davis and Reuben Hersh, *The Mathematical Experience* (Cambridge, Mass.: Birkhäuser Boston, 1981), 217–18.

126 *Known as the parallel postulate:* This is actually a version put forth by John Playfair in 1795. Euclid's original fifth postulate stated the mathematical equivalent, that if a straight line crosses two straight lines and the interior angles on the same side add to less than 180 degrees, the two straight lines will eventually meet on that side of the straight line [Morris Kline, *Mathematics and the Search for Knowledge* (New York: Pantheon, 1985), 150].

129 *"Lincoln's frank unwillingness":* David Berlinski, *A Tour of the Calculus* (1995), 199 n. 1. See also Cohen, *Science and the Founding Fathers,* 57 (describing the Declaration of Independence as deductive and the Gettysburg Address as inductive).

129 *The study of other countries' constitutional systems:* See *Printz v. U.S.,* 521 U.S. 898, 977 (Justice Breyer dissenting) (stating that "there may be relevant political and structural differences between their systems and our own. But their experience may nonetheless cast an empirical light on the consequences of different solutions to a common legal problem . . .").

129 Miami Herald Publishing: 418 U.S. 241 (1974).

129 *In England:* See, for example, *Lee v. Bude and Torrington Ry. Co.,* L.R. 6 C.P. 576, per Wiles, 582 (1871).

130 Marbury: 5 U.S. (1 Cranch) 137 (1803).

130 *Even if the results flowed logically:* As Morris Kline explained, "[T]he legal right of the individual to engage in private enterprise is a principle in a capitalistic system of government, just as the Euclidean parallel axiom is an axiom in the Euclidean system of geometry. The differences among fascist, democratic, and communistic forms of government stem from differences in fundamental principles, just as the differing theorems in the several geometries stem from different axioms. And just as each geometry attempts to treat physical space, each political system attempts to treat the social order" (Kline, *Mathematics in Western Culture,* 464).

130 *"Paradoxically, although the new geometries impugned":* Kline, *Mathematics in Western Culture,* 430. See, for example, Burton M. Leiser, "Threats to Academic Freedom and Tenure," *Pace Law Review* 15 (1994): 15 (describing the need to protect the right to express a variety of thoughts).

130 *"[I]t appears to me":* Thomas Jefferson, "Notes on the State of Virginia," 1787; reprinted in *Jefferson: Writings,* 123, 206.

130 *"one universal Father":* Letter from Benjamin Banneker to Thomas Jefferson, August 19, 1791, reprinted in Silvio A. Bedini, *The Life of Benjamin Banneker* (New York: Scribner, 1972) 152–56.

131 *"I am happy to inform you":* Letter from Thomas Jefferson to the Marquis de Condorcet, August 30, 1791, reprinted ibid., 159. Interestingly, Benjamin Franklin had not needed such evidence. He had also written Condorcet, but in 1774, saying that African Americans were not "deficient in natural Understanding . . . [but] they have not the Advantage of Education" (Cohen, *Science and the Founding Fathers,* 193).

132 *"whatever shape you design":* Osserman, *Poetry of the Universe,* 164.

132 *"sphere of jurisdiction":* Federalist No. 45. See also Federalist No. 39 (stating that "the local or municipal authorities form distinct and independent portions of the supremacy, no more subject, within their respective spheres, to the general authority, than the general authority is subject to them, within its own sphere"). See further Tribe, "Taking Text and Structure Seriously," 1248 ("[C]onstitutional topology counsels against the . . . error of ignoring how the surfaces and edges of a complex structure connect and intersect").

132 *"a municipality is merely": United Building and Construction Trades Council v. Camden,* 465 U.S. 208, 215 (1984).

133 *Article V gives absolutely no role:* For a fuller discussion of this point, see Chapter 7.

133 *"sole Power of Impeachment":* U.S. Const., Art. I, §2, cl. 5, and §3, cl. 7. See *Nixon v. U.S.,* 506 U.S. 224 (1993) (holding that the Senate could define the meaning of the word *try* in adjudicating impeachment proceedings).

133 *"the Chief Justice shall preside":* U.S. Const., Art. I, §3, cl. 7.

134 *"Constitutional Topology":* The title of this section is taken from Tribe, "Taking Text and Structure Seriously," 1245. This ground-breaking article was the first to use topology to explore the structure of the Constitution.

135 *"deeper invariances":* Paulos, *Beyond Numeracy,* 235.

135 *all topological figures are flexible:* Casti, *Five Golden Rules,* 50.

135 *The number of holes:* Another topological property is the number of edges, or boundaries, a shape has. A circle has no edge, while a straight line has two endpoints. Thus, one cannot topologically transform a circle into a line.

135 *One of the fundamental theorems of topology:* Technically, this does not apply to a special category, called nonorientable surfaces, which are

those surfaces, like the Möbius strip, on which one cannot distinguish right-handness from left-handness, or clockwise from counterclockwise.

136 *"framework has been sufficiently flexible": New York v. U.S.,* 505 U.S. 144, 157 (1992).

136 *"the power of the Federal government":* Ibid.

136 *some changes can be seen as merely stretching:* Tribe, "Taking Text and Structure Seriously," 1237.

136 *"the unique contribution": U.S. v. Lopez,* 514 U.S. 549, 575 (1995) (Justice Kennedy concurring).

137 *"In the compound republic of America":* Federalist No. 51 (Madison).

137 *in 1895, the Court held: U.S. v. E.C. Knight Co.,* 156 U.S. 1 (1895).

137 *By 1942, however, the Court: Wickard v. Filburn,* 317 U.S. 111 (1942).

137 *This expansion of congressional power:* See, for example, *Lopez,* 514 U.S. 574 (Justice Kennedy concurring).

137 *Yet, in 1995, the Court ruled: U.S. v. Lopez,* 514 U.S. 549 (1995).

137 *"that there never will be":* Ibid., 567–68.

138 *the Court in 1985 upheld: Garcia v. San Antonio Metropolitan Transit Authority,* 469 U.S. 528 (1985), overruling *National League of Cities v. Usery,* 426 U.S. 833 (1976).

138 *"the States as* States *have legitimate interests": Garcia,* 469 U.S. 528 (Justice O'Connor dissenting).

138 *the Court subsequently has held:* See *New York v. U.S.,* 505 U.S. 144 (1992); *Printz v. U.S.,* 521 U.S. 898.

138 *"reducing them to puppets": Printz,* 512 U.S. 928 [quoting *Brown v. EPA,* 512 F. 2d 827, 839 (9th Cir. 1975), vacated as moot, 431 U.S. 99 (1977)].

138 *"our constitutional system of dual sovereignty":* Ibid., 935.

138 Chadha: 462 U.S. 919 (1983). In analyzing the legislative veto, Professor Laurence Tribe analogized Congress's legislative authority to a "roughly spherical solid" and said that if Congress were able to veto executive actions, it would be akin to taking a piece of the solid, stretching it out, and then looping it back into the original solid at another point: "We would have created a hole in the figure, giving it at least some of the properties we would expect to see in a doughnut, rather than in a sphere. This was the type of transformation effected by the legislative veto, which represented not simply a bump or extension of normal delegation, but a loop back into congressional power." Professor Tribe concluded, to permit the legislative veto would be to "fundamentally alter the topology of the government framework the Constitution describes" (Tribe, "Taking Text and Structure Seriously," 1238).

139 *the seven bridges of Königsberg:* For an excellent discussion of the solution to the problem of the bridges of Königsberg, see Hoffman, *Archimedes' Revenge,* 155–58.

140 *In 1905, Albert Einstein proved:* This was part of Einstein's special theory of relativity. An English translation of the original paper, entitled "On the Electrodynamics of Moving Bodies," can be found in Hendrik A. Lorentz, Albert Einstein, Hermann Minkowski, and Herman Weyl, *The Principle of Relativity: A Collection of Original Memoirs on the Special and General Theory of Relativity* (New York: Dover, 1952), 36–65. Einstein wrote a popular exposition of his theory in Albert Einstein, *Relativity* (15th ed.; New York: Crown, 1961), 25–57.

140 *floor of a train moving at extraordinarily high speed:* To make a noticeable difference, we are imagining an inhuman speed of more than three-fourths of the speed of light, or over 600 million miles per hour. Also, the numbers used to estimate the observer's view of distance are for illustrative purposes; they have not been calibrated to any particular speed of the train.

141 *It is literally meaningless:* See Kline, *Mathematics in Western Culture,* 433.

142 Plessy: 163 U.S. 537 (1896). For an excellent discussion of *Plessy,* see Higginbotham, *Shades of Freedom,* 111–15.

142 *"If this be so": Plessy,* 163 U.S. 551.

142 *"Every one knows":* Ibid., 5517–60 (Justice Harlan dissenting).

142 *"at the very least, prohibits government": County of Allegheny v. A.C.L.U.,* 492 U.S. 573, 593–94 (1989), quoting *Lynch v. Donnelly,* 465 U.S. 668, 687 (1984) (Justice O'Connor concurring).

143 *"is sufficiently likely to be perceived": County of Allegheny,* 492 U.S. 594, quoting *Grand Rapids School District v. Ball,* 473 U.S. 373, 390 (1985). The specific ruling of *Grand Rapids,* prohibiting public paid teachers from going into religious school for remedial education, was overturned in 1997.

143 Lynch: 465 U.S. 668 (1984).

143 *"The display engenders a friendly community spirit":* Ibid., 685.

143 *"Christians feel constrained":* Ibid., 727 (Justice Blackmun dissenting.) The placement of a nativity scene in a county courthouse, without the nonreligious elements, was held unconstitutional in *County of Allegheny v. A.C.L.U.,* 492 U.S. 573 (1989).

144 *"What to most believers": Lee v. Weisman,* 505 U.S. 577, 592 (1992). See also *Engle v. Vitale,* 370 U.S. 421, 431 (1962) (stating, "When the power, prestige, and financial support of government is placed behind a particular religious belief, the indirect coercive pressure upon religious

minorities to conform to the prevailing officially approved religion is plain").

144 *the Court has chosen to rule:* See, for example, *Engle,* 370 U.S. 421.

144 *through equations known as Lorentz transformations:* The equations require that we know the other person's velocity, and the time and distance associated with our own point of reference. See generally Delo E. Mook and Thomas Vargish, *Inside Relativity,* (Princeton: Princeton University Press, 1987), 96–102. The Lorentz transformations are named for physicist Hedrik A. Lorentz, who devised them before Einstein published his theory of relativity.

144 Capitol Square Review: 515 U.S. 753 (1995).

145 "*Private religious speech*": Ibid., 766 (Justice Scalia, plurality opinion).

145 "*a hypothetical observer*": Ibid., 780 (Justice O'Connor concurring).

145 "*It is especially important to take account*": Ibid., 799 (Justice Stevens dissenting).

145 "*A person who views an exotic cow*": Ibid., 800 n. 5 (Justice Stevens dissenting).

145 Mozert: 827 F. 2d 1058, 1074 (6th Cir. 1987) (Justice Boggs concurring), *cert. denied* 484 U.S. 1066 (1988).

146 "*While many of the passages*": *Mozert,* 827 F. 2d 1069 (emphasis added).

146 "*[T]he school board is indeed entitled*": Ibid., 1074 (Justice Boggs concurring).

146 "*a reachable perception that she is being forced*": *Lee v. Weisman,* 505 U.S. 577, 593 (1992).

146 "*the Constitution forbids the State*": Ibid., 596.

146 *when the school authorities disagree with nonfundamentalist students: Mozert,* 827 F. 2d 1081 (Justice Boggs concurring). It also certainly should not be that an interest in "promoting cohesion among a heterogenous democratic people" [Ibid., 1072 (Justice Kennedy concurring), quoting *McCollum v. Board of Education,* 333 U.S. 203, 216 (1948) (Justice Frankfurter concurring)] is served by inflicting offensive texts on religious students but not by inflicting prayers on those who are nonreligious.

147 "*would result in substantial disruption*": *Mozert,* 827 F. 2d 1072 (Justice Kennedy concurring).

147 *discount either student's perspective as unreasonable:* In the words of the Supreme Court, "Religious beliefs need not to be acceptable, logical, consistent, or comprehensible to others in order to merit First Amendment protection" [*Thomas v. Review Board,* 450 U.S. 707, 714 (1981)].

147 *"Every closed curve in the plane":* Kasner and Newman, *Mathematics and the Imagination,* 276. This theorem was first stated by a French mathematician, Camille Jordan.

147 *Much of the history of our Constitution:* As Justice Ginsburg has written, "A prime part of the history of our Constitution . . . is the story of the extension of constitutional rights to people once ignored or excluded" [*U.S. v. Virginia,* 518 U.S. 515, 557 (1996)].

147 *"this is a Christian nation": Church of Holy Trinity v. U.S.,* 143 U.S. 457, 471 (1892).

147 *"racial instincts": Plessy,* 163 U.S. 551.

147 *"He drew a circle":* E. Markham, "Outwitted," in *Best Loved Poems of the American People,* ed. Hazel Fellerman (Garden City, N.Y.: Garden City Books, 1957), 67.

## 8. Infinity and the Constitution

148 *"Just as sight recognizes darkness":* Proclus, quoted in Ray Hemmings and Dick Tahta, *Images of Infinity* (Stradbroke, England: Tarquin, 1984), 15.

148 *There was once a very popular hotel:* Ibid., 82–83.

150 *counting is really the same:* Kasner and Newman, *Mathematics and the Imagination,* 28–29 ("Learning to compare is learning to count. Without knowing anything about numbers, one may ascertain whether two classes have the same number of elements; for example, barring prior mishaps, it is easy to show that we have the same number of fingers on both hands by simply matching finger with finger on each hand").

150 *He then described all infinite sets:* Hemmings and Tahta, *Images of Infinity,* 90.

151 *"whose elements can be paired off":* Ibid., 88. In 1872, a German mathematician, Richard Dedekind, wrote more formally that "A system S is said to be infinite when it is similar to a proper part of itself" (ibid.).

151 *"[I]f we accept one-to-one correspondence":* Kline, *Mathematics in Western Culture,* 399–400.

151 *Cantor designated the countable sets:* Eli Maor, *To Infinity and Beyond* (Princeton, N.J.: Princeton University Press, 1987), 60. The symbol ℵ (pronounced "aleph") is the first letter of the Hebrew alphabet.

152 *decimals can be either:* Terminating decimals are just a special form of repeating decimals, since 0.123 is the same as 0.123000. . . .

152 *Cantor showed that no matter:* To see an example of Cantor's proof, con-

sider one such listing of random decimal numbers (derived from Hemmings and Tahta, *Images of Infinity,* 97).

$$\begin{array}{ccccccc} 0 & .6 & 1 & 1 & 5 & 5 & \ldots \\ 0 & .5 & 1 & 2 & 8 & 4 & \ldots \\ 0 & .3 & 5 & 8 & 6 & 0 & \ldots \\ 0 & .2 & 2 & 3 & 5 & 4 & \ldots \\ . & . & . & . & . & . & . \end{array}$$

Next, take the sequence of numbers starting at the top left of our list and work diagonally down, so we have the numbers 6 1 8 5. . . . Then, create a new decimal by adding 1 to each of these numbers and get 0.7296. . . .

We know that this number is not on our list of decimals. By the way we created it, it must differ from every number on the list by at least one digit. Because we can repeat this process indefinitely, no listing of decimals will ever be orderly enough to count. Decimals, the real numbers, are not countable.

Note that for this example, any listing of decimals could have been used, or, even a generalized notation such as

$$\begin{array}{ccccc} A_1, & A_2, & A_3, & A_4, & \ldots \\ B_1, & B_2, & B_3, & B_4, & \ldots \\ . & . & . & . & \ldots \end{array}$$

152 *So, even though:* The power of the set of real numbers, *C,* is frequently referred to as $\aleph_1$, so we can say that $\aleph_1 > \aleph_0$. There are sets with more members then *C* (or $\aleph_1$), and if one desires, one can set up "an infinite hierarchy of infinite sets" (Maor, *To Infinity and Beyond,* 64).

152 *there are three relevant points about infinity:* Some authors on transfinite arithmetic have been almost defensive about their readers' possible reactions to the topic:

*It has been suggested that at this point the tired reader puts the book down with a sigh—and goes to the movies. We can only offer in mitigation that this proof . . . is tough and no bones about it. You may grit your teeth and try to get what you can out of them, or conveniently omit them. The essential thing to come away with is that Cantor found that the rational fractions are countable but that the set of real numbers is not. Thus, in spite of what common sense tells you, there are no more fractions than there are integers and there are more real numbers between 0 and 1 than there are elements in the whole class of integers.*

—*Kasner and Newman,* Mathematics and the Imagination, *48–49*

152 *"From generation to generation":* Letter to Samuel Kercheval, July 12, 1816, in *Jefferson: Writings,* 1402.

152 *"[N]o society can make a perpetual constitution":* Ibid., 959, 963.

152 *"[T]his Constitution is not framed to answer":* Charles Pinckney speaking at the South Carolina Ratification Convention, May 20, 1788, in *Founders' Constitution,* vol. 3, Kurland and Lerner, 396. Similar sentiments were expressed by Edward Carrington: "The prevailing impression as well in, as out of, Convention, is, that a federal Government adapted to the permanent circumstances of the Country, without respect to the habits of the day, be formed . . ." (letter from Edward Carrington to Thomas Jefferson, June 9, 1787, in *Founders' Constitution,* vol. 1, Kurland and Lerner, 252).

153 *"All the descendants":* Henry Campbell Black, *Black's Law Dictionary,* 1050 (5th ed.; St. Paul: West, 1979).

153 *"the Articles of this confederation":* Art. of Conf., Art. XIII.

153 *"Note that it was":* In fact, when Congress authorized the convention for "revising the articles of confederation" (which, of course, ended up drafting the brand new Constitution), the authorization began, "WHEREAS, There is provision in the articles of Confederation and perpetual Union, for making alterations therein . . ." [Federalist No. 40 (Madison)].

153 *"[T]he very attempt to make* perpetual *constitutions":* Giles Hickory [Noah Webster], "On the Bill of Rights," *American Magazine* 1 (December 1787): 13–15, quoted in Gordon S. Wood, *The Creation of the American Republic: 1776–1787* (1969), 379.

153 *Even Thomas Jefferson:* Jefferson was actually a somewhat ambivalent supporter of the Constitution, as can be seen in the following letter: "I wish with all my soul, that the nine first conventions may accept the new constitution, because this will secure to us the good it contains, which I think great and important. But I equally wish, that the four latest conventions, which ever they may be, may refuse to accede to it, till a declaration of rights be annexed" (Letter to Alexander Donald, February 7, 1788, in *Jefferson: Writings,* 919).

153 *"Whether one generation of men has a right to bind":* Letter to James Madison, September 6, 1789, in *Jefferson: Writings,* 959.

154 *"Every constitution then":* Ibid., 963. According to Jefferson's letter, the

median lifespan for those 21 years and older was "18 years and 8 months, or say 19 years as the nearest integral number" (ibid).

154 *He argued that, because each generation should be free:* Letter to Samuel Kercheval, July 12, 1816, in *Jefferson: Writings,* 1402.

154 *"[A] solemn opportunity of doing this":* Ibid.

154 *"The plan now to be formed":* George Mason, speaking at the Federal Convention, June 11, 1787, in *Founders' Constitution,* vol. 4, Kurland and Lerner, 576.

155 *"The Constitution of any government":* James Iredell, speaking at the North Carolina Ratification Convention, July 29, 1788, ibid., 582.

155 *"Happy this, the country we live in!":* Ibid.

155 *"Until the people have, by some solemn and authoritative act":* Federalist No. 78 (Hamilton).

156 Nebraska Press: 427 U.S. 539 (1976). A local judge hearing a case involving a particularly gruesome murder had issued an order restraining news reporting which discussed confessions or other evidence "strongly implicative" of the accused.

156 *"The Liberty of the Press":* Argus, *Providence U.S. Chronicle,* November 8, 1787, in *Documentary History,* 320–21, quoted in *McIntyre v. Ohio Elections Commission,* 514 U.S. 334, 366 (1995).

156 *"England, from whom the Western World": Irvin v. Dowd,* 366 U.S. 717, 721 (1961).

156 *"authors of the Bill of Rights did not undertake": Nebraska Press,* 427 U.S. 561.

157 *"there is an inherent conflict":* Ibid., 618 (Justice Brennan concurring).

157 *"[E]very moment's continuance": New York Times Co. v. U.S.,* 403 U.S. 713, 714–15 (1971) (Justice Black concurring). The dissenting justices argued that before the case was finally decided, a trial should have been held, "permitting the orderly presentation of evidence from both sides" [ibid., 761–62 (Justice Blackmun dissenting)].

157 *"The Federal Government is without any power": Ginzburg v. U.S.,* 383 U.S. 463, 476 (1966) (Justice Black dissenting).

157 *Instead, there is general acceptance of the principle:* In *Near v. Minnesota,* 283 U.S. 697, 716 (1931), the Court stated that a prior restraint could be imposed on a newspaper to prevent the "actual obstruction to [the Government's] recruiting service or the publication of the sailing dates of transports or the number and location of troops."

158 *"the gravity of the 'evil'": Dennis v. U.S.,* 183 F. 2d 201, 212 (1950), aff'd. 341 U.S. 494 (1951).

158 *a restriction will be viewed as constitutional:* Judge Richard Posner has

described Judge Learned Hand's test as "[a]n economic formula," where *B* is "the cost of the reduction in the stock of ideas as a result of the government's actions" [Richard Posner, *Economic Analysis of Law* (4th ed.; Boston: Little, Brown, 1992), 667]. In economic terms, if *B* is less than *PL*, "it is efficient for the Government to take steps against the speaker" (Ibid.).

158 *This step was missed by the U.S. district court judge: U.S. v. Progressive, Inc.*, 467 F. Supp. 990 (W.D. Wis.), appeal dismissed, 610 F. 2d 819 (7th Cir. 1979).

159 *the least restrictive necessary:* See, for example, *City of Houston v. Hill*, 482 U.S. 451, 465 (1987) (requiring regulation of speech to be "narrowly tailored").

160 *"cannot reasonably and objectively be regarded": Webster v. Reproductive Health Services*, 492 U.S. 490 (1989) (Justice Blackmun dissenting) (emphasis added).

160 *"The Court apparently values* the conveniences": Justice White's dissent in *Roe v. Wade*, appears in *Doe v. Bolton*, 410 U.S. 179, 222 (1973) (Justice White dissenting).

161 *"most basic decisions": Planned Parenthood of Southeastern Pennsylvania v. Casey*, 505 U.S. 833, 849 (1992). The terms *pro-life* and *pro-choice* are political terms that should not be used to characterize the constitutional position of the justices. They are, however, a convenient short way of describing the opposing sides in the debate.

161 *"[I]n the balance":* Ruth Bader Ginsburg, "Some Thoughts on Autonomy and Equality in Relation to *Roe v. Wade*," *North Carolina Law Review* 63 (1985): 375, 383. In *Casey*, the Court stated that "The mother who carries a child to full term is subject to anxieties, to physical restraints, to pain that only she must bear. . . . Her suffering is too intimate and personal for the state to insist, *without more*, upon its own vision of the woman's role" (505 U.S. 852) (emphasis added).

## 9. The Incomplete Constitution

162 *"Our world is endlessly more complicated":* Rudy Rucker, *Mind Tools*, 247.

162 *Russell's Barber:* This story is derived from Bertrand Russell's Paradox of the Barber [see Morris Kline, *Mathematics: The Loss of Certainty* (New York: Oxford University Press, 1980), 205. A version of this paradox also appears in Maor, *To Infinity and Beyond*, 255].

164 *A metamathematical statement:* Casti, *Five Golden Rules,* 155.

164 *"We now come to the goal":* Kurt Gödel, *On Formally Undecidable Propositions of Principa Mathematica and Related Systems I,* trans. Jean van Heijenoort, ed. S. G. Shanker, in *Gödel's Theorem in Focus* (London: Routledge, 1988), 30. This sentence has also been translated, "We now come to the object of our exercises" [see Kurt Gödel, *On Formally Undecidable Propositions of Principa Mathematica and Related Systems I,* trans. B. Meltzer (1931; New York: Dover, 1992), 56].

165 *What gives the statement the power to transcend:* This distinction is part of the fundamental difference between semantics (whether a statement's meaning is true) and syntax (whether under the "grammar" of the system a given statement is a theorem). See, for example, Mike Townsend, "Implications of Foundational Crises in Mathematics: A Case Study in Interdisciplinary Research, *Washington Law Review* 71 (1996): 51, 118–19.

165 *the formal system is destined to remain incomplete:* Gödel also proved a second incompleteness theorem, establishing that a formal system could not prove itself consistent.

166 *"It would be a large mistake":* Douglas Hofstadter, *Gödel, Escher, Bach: An Eternal Golden Braid* (New York: Basic Books, 1979), 696.

166 *"one would have to provide":* Townsend, "Implications of Foundational Crises in Mathematics," 130. See also Kevin W. Saunders, "Realism, Ratiocination, and Rules," *Oklahoma Law Review* 46 (1993): 219, 220: "In order for Gödel's Theorem to apply to law[,] . . . [more] is required than showing that laws may be self-referential or that law may have rules and metarules. Required is a demonstration that the metalanguage of law—legal English—can in some sense be embedded in the law."

166 *It also would be a mistake to treat Gödel's theorem:* For an impressive review of the varied attempts by legal scholars to apply Gödel's theorem to law, see Townsend, "Implications of Foundational Crises in Mathematics," 117–35. Some have argued that the indeterminacy of law is traceable to the unscientific nature of law, that there is no one overarching universal legal truth. See, for example, Dow, "Gödel and Langdell," 707, 716–20.

167 *the passive virtues:* See generally, Alexander Bickel, *The Least Dangerous Branch* (Indianapolis: Bobbs-Merrill, 1965), 111.

167 *"will not pass upon": U.S. v. Hayman,* 342 U.S. 205, 223 (1952). See also *Ashwander v. Tennessee Valley Authority,* 297 U.S. 288, 348 (1936) (Justice Brandeis concurring): "When the validity of an act of Congress is

drawn in question, and even if a serious doubt of constitutionality is raised, it is a cardinal principle that this Court will first ascertain whether a construction of a statute is fairly possible by which the question may be avoided."

167 *When the Supreme Court interprets a statute narrowly:* See, for example, *Hamling v. U.S.,* 418 U.S. 87, 113 (1974) (interpreting the statutory ban on mailing "obscene, lewd, indecent, filthy, or vile" material so as to bar the mailing of only legally obscene material).

167 *"an insubstantial state issue": Mayor of Philadelphia v. Educational Equality League,* 415 U.S. 605, 628–29 (1974). See also *Siller v. Louisville & N.R.R. Co.,* 213 U.S. 175, 193 (1909) (stating that the metarule of avoiding constitutional decisions is "not departed from without important reasons").

167 Marbury: 5 U.S. 137 (1804).

168 McCulloch: 17 U.S. 316 (1819).

169 U.S. v. Nixon: 418 U.S. 683 (1974).

169 Bush v. Gore: 531 U.S. 98 (2000).

170 *"An analogous metaissue":* Compare *Dellums v. Bush,* 752 F. Supp. 1141 (D.D.C. 1990) (finding the issue unripe but stating that an invasion of Kuwait by "several hundred thousand U.S. servicemen . . . could be described as a 'war'").

170 *While the President:* Compare U.S. Const., Art. II. §2 with Art. I, §8.

171 *Take the issue of whether a constitutional decision:* The paradoxical nature of this issue is well analyzed in George P. Fletcher, "Paradoxes in Legal Thought," *Columbia Law Review* 85 (1985): 1263, 1276–77.

171 Miranda: 384 U.S. 436 (1966).

171 *"available only to persons": Johnson v. New Jersey,* 384 U.S. 719, 734 (1966).

171 *At first, the Court's way out of this dilemma:* See, for example, *Desist v. United States,* 394 U.S. 244, 246 (1969), in which the Court stated, "We have concluded, however, that to the extent *Katz* [*Katz v. U.S.,* 389 U.S. 347 (1967)] departed from previous holdings of this Court, it should be given wholly prospective application."

171 *Later, the Court decided: Shea v. Louisiana,* 470 U.S. 51, 59 (1985), accord *U.S. v. Johnson,* 457 U.S. 537, 548 (1982).

171 "stare decisis *ought to be applied": Planned Parenthood of Southeastern Pennsylvania v. Casey,* 505 U.S. 833, 993 (Justice Scalia dissenting).

172 *"The doctrine of precedent cannot be authoritatively":* J. C. Hicks, "The Liar Paradox in Legal Reasoning," *Cambridge Law Journal,* 29 (1971): 275, 284.

172 *44 Liquourmart:* 517 U.S. 484, 510 (1996) (Justice Stevens, plurality

opinion). For example, Justice Stevens stated that "Because the 5-4 decision in [*Posadas de Puerto Rico Associates v Tourism Co.,* 478 U.S. 328 (1986)] marked such a sharp break from our prior precedent[,] . . . we decline to give force to its highly deferential approach."

172 *But if you believe in equal deference:* See also Hicks, "The Liar Paradox in Legal Reasoning," 288 (stating that "a court can without illogicality change from the rejection to the acceptance of *stare decis* but cannot do the reverse").

172 *the House of Representatives voted to require:* See Julius J. Marke, *Vignettes of Legal History* (South Hackensack, N.J.: F. B. Rothman, 1965), 149–51.

172 *But what if it had become law?*: The following scenario is courtesy of Douglas Hofstadter, *Mathematical Themas* (New York: Basic Books, 1985), 71–72.

173 *"the federal judiciary is supreme": Cooper v. Aaron,* 358 U.S. 1, 18 (1958).

173 "Marbury *put a firm halt":* Dow, "Gödel and Langdell," 707, 720–21.

173 *"logically antecedent to the written constitution":* Larry Alexander and Frederick Schauer, "On Extrajudicial Constitutional Interpretation," *Harvard Law Review* 110 (1997): 1359, 1369.

173 *"it is even clearer that such a specification":* Ibid., 1370.

173 *The Constitution prohibits amendments:* U.S. Const., Art. V.

173 *Just before the Civil War:* The Corwin Amendment stated: "No amendment shall be made to the Constitution which will authorize or give to Congress the power to abolish or interfere, within any State, with the domestic institutions thereof, including that of persons held to labor or service by the laws of said State," Joint Resolution 13, 36th Congress, 2d session, 12 Stat. 251; Congressional Globe, 36th Congress, 2d session, 1263 (1861).

173 *The Amendment only garnered two state ratifications:* For an excellent discussion of this provision, see Michael Stokes Paulsen, "A General Theory of Article V: The Constitutional Lessons of the Twenty-seventh Amendment, *Yale Law Journal* 103 (1993): 677, 698–99.

174 *Mathematically, the theory of types:* For discussions of the theory of types, see Paulos, *Beyond Numeracy,* 215–16; Devlin, *Mathematics,* 58-60; Herbert B. Enderton, *Elements of Set Theory* (New York: Academic Press, 1977), 6–22.

174 *Thus, a statute can bar the change of a regulation:* For example, once Congress statutorily affirmed the Federal Communications Commission's rules requiring cable television systems to carry local broadcast signals, the "must-carry" rules, the FCC was barred from

repealing its rules [see *Turner Broadcasting Systems, Inc. v. FCC,* 512 U.S. 622 (1994)].

175 *a constitutional provision can prevent the changing of a statute:* The Twenty-sixth Amendment, guaranteeing 18-year-olds the right to vote, prevents Congress from repealing its law granting 18-year-olds the right to vote. See generally *Oregon v. Mitchell,* 400 U.S. 112 (1970) (upholding a federal law granting 18-year-olds the right to vote in federal elections but striking that portion which granted 18-year-olds the right to vote in state and local elections).

175 Morrison: 487 U.S. 654 (1988).

176 *This is a dangerous rationale:* See, for example, Stephen Carter, "The Independent Counsel Mess," *Harvard Law Review* 102 (1988): 105, 134–35.

176 *Douglas Hofstadter had presciently foreseen:* Hofstadter, *Gödel, Escher, Bach,* 692–93. Hofstadter's example of a "strange loop in government" was narrower, focusing on the idea of who will police the police: "Other curious tangles which arise in government include the FBI investigating its own wrongdoings. . . ."

177 *"the separation of powers may prevent us": Morrison,* 487 U.S. 710 (Justice Scalia dissenting).

178 *City of Boerne:* 521 U.S. 507 (1997).

178 *Employment Division, Department of Human Resources v. Smith:* 494 U.S. 872 (1990).

179 *The Supreme Court refused to address: Nixon v. U.S.,* 506 U.S. 224 (1993).

179 *"be Judge of the Elections, Returns and Qualifications":* U.S. Const., Art. I, § 5, cl. 1.

179 *When the House tried to bar Representative Adam Clayton Powell: Powell v. McCormack,* 395 U.S. 486 (1969).

179 *"vesting an improper & dangerous power":* James Madison, August 10, 1787, in *Debates in the Federal Convention,* 374.

179 *Should a majority of the Legislature be composed":* Hugh Williamson, August 10, 1787, in *Debates in the Federal Convention,* 375.

179 *In theory, Congress will be reluctant:* "Because legislators are unable to exempt themselves from law execution, they must, in enacting laws, take the perspective of ordinary citizens subject to the force of the law" [Cass R. Sunstein, "Constitutionalism After the New Deal," *Harvard Law Review* 101 (1987): 421, 434]. See also David Epstein, *The Political Theory of the Federalist* (Chicago: University of Chicago Press, 1984), 129–30.

180 *"too great a temptation":* John Locke, *The Second Treatise of Government*

(1689), ed. Peter Laslett (New York: N. A. L., 1965), sec. 143.

180 *the Congressional Accountability Act:* Public Law 104-1, 109 Stat. 3 (1995).

180 *the most recent amendment to the Constitution:* "No law, varying the compensation for the services of Senators and Representatives, shall take effect, until an election of Representatives shall have intervened" (U.S. Const. Amend. XXVII).

181 *Officially, the formal determination of the validity of the amendment:* U.S. Code, vol. 1, sec. 106b (1988) provides: "Whenever official notice is received at the National Archives and Records Administration that any amendment proposed to the Constitution of the United States has been adopted, according to the provisions of the Constitution, the Archivist of the United States shall forthwith cause the amendment to be published, with his certificate, specifying the States by which the same may have been adopted, and that the same has become valid, to all intents and purposes, as a part of the Constitution of the United States."

181 *determining whether the appropriate proceedings were followed:* See, for example, *Clinton v. City of New York,* 524 U.S. 417 (1998), in which the Court ruled that the procedures provided by the Line Item Veto Act for enacting a law violated the Presentment Clause of the Constitution, Art. I, § 7, cl. 2.

181 Coleman: 307 U.S. 433 (1939).

181 *the Child Labor Amendment:* The proposed amendment stated that "Congress shall have power to limit, regulate, and prohibit the labor of persons under eighteen years of age" [H.R. J. Res. 184, 68th Cong., 1st sess., 43 Stat. 670 (1924)]. The amendment represented an attempt to reverse the rulings of the Court in *Hammer v. Dagenhart,* 247 U.S. 251 (1918) and *Bailey v. Drexel Furniture Co.,* 259 U.S. 20 (1922).

181 *as lacking a majority opinion:* Paulsen, "A General Theory of Article V," 677, 717.

181 *Congress has the ultimate authority:* See, for example, *Goldwater v. Carter,* 444 U.S. 996, 1002–3 (1979) (Justice Rehnquist concurring).

181 *"The proposed constitutional amendment at issue":* Ibid., 1001 (Justice Powell concurring).

182 *Amendments also have been designed:* The Twenty-fifth Amendment provides for the removal of a sitting president, and the Seventeenth Amendment changed the way senators were elected. See generally, Walter Dellinger, "The Legitimacy of Constitutional Change: Rethinking the Amending Process," *Harvard Law Review* 97 (1983): 380, 414–16.

182 *judicial review of the amending process is only:* "One can fairly debate whether such judicial abstention is ever legitimate. But it clearly cannot be justified when the proposed amendment gores Congress' ox, not the Court's . . ." [Paulsen, "A General Theory of Article V," 677, 717].

182 *But what if Congress proposed:* Many amendments to the Constitution have increased congressional power by granting Congress the power to enforce the specific amendment. See, for example, U.S. Const., Amend. XIII, §2; Amend. XIV, §5; Amend. XV, §2; Amend. XIX, §2; Amend. XXIII, §2; Amend. XXIV, §2; Amend. XXVI, §2. And, of course, the Sixteenth Amendment gave Congress the power to levy the income tax.

182 *what if one Congress wisely proposed:* See, for example, U.S. Const., Amend. XVII, limiting the ability of Congress to raise Congressional salaries. See also U.S. Const., Amend. XVII, providing for popular election of senators, rather than selection by state legislatures.

182 *"It would be improper to require":* George Mason, June 11, 1787, in *Debates in the Federal Convention,* 90. In response to such concerns, Article V includes a method for a state to bypass congressional discretion by calling for a new constitutional convention. See also Federalist No. 43 (Madison): "It, moreover, equally enables the general and the State governments to originate the amendment of errors, as they may be pointed out by the experience on one side, or on the other."

183 *"[I]t may well be the path of practical wisdom":* Hicks, "The Liar Paradox in Legal Reasoning," 291. There have been several interesting articles on legal lessons that can be learned from both the Incompleteness Theorem and self-reference in general. See, for example, John M. Rogers and Robert E. Molzon, "Some Lessons About the Law from Self-Referential Problems in Mathematics," *Michigan Law Review* 90 (1992): 992; Mark R. Brown and Andrew C. Greenberg, "On Formally Undecidable Propositions of Law: Legal Indeterminacy and the Implications of Metamathematics," *Hastings Law Journal* 43 (1992): 1439; Dow, "Gödel and Langdell," 707; Stuart Banner, "Please Don't Read the Title," *Ohio State Law Journal* 50 (1989): 243; Steven P. Goldberg, "On Legal and Mathematical Reasoning," *Jurimetrics* 22 (1981): 83; Susan K. Houser, "Metaethics and the Overlapping Consensus," *Ohio State Law Journal* 54 (1993): 1139.

184 *shortly thereafter, Gödel became a citizen:* The anecdote of Gödel's citizenship exam is taken from Peter Suber, *The Paradox of Self-Amendment* (New York: Peter Lang, 1990), 212; Ed Regis, *Who Got Einstein's Office?* (Reading, Mass.: Addison-Wesley, 1987), 57–58; Solomon Feferman, "Gödel's Life and Work," in *Kurt Gödel: Collected*

*Works,* vol. 1 (New York: Oxford University Press, 1986), 12. While it is agreed that Gödel had discovered a flaw in the Constitution, it is not clear that the flaw he discovered was the breadth of permitted amendments.

## 10. Constitutional Chaos

185 *"A very small cause":* Peterson, *Islands of Truth,* 261.

185 *Unfortunately, that intuition leaves a person:* Robert May, "Simple Mathematical Models with Very Complicated Dynamics," *Nature* 262 (1976): 459, 467 (stating, in his criticism of the focus on linear thinking in mathematics courses, "The mathematical intuition so developed ill equips the student to confront the bizarre behavior exhibited by the simplest of discrete nonlinear systems. . . . Yet such nonlinear systems are surely the rule, not the exception . . .").

186 *functions of this sort are called linear functions:* A linear function has the form $f(x) = mx + b$, where $m$ and $b$ are each a real number.

187 *we do not live in a linear world:* Many of the examples given are derived from Casti, *Five Golden Rules,* 203, and James Gleick, *Chaos* (New York: Viking Penguin, 1987), 24.

187 *Up until the 1960s:* See, for example, Thomas P. Dick and Charles M. Patton, *Calculus of a Single Variable* (Boston: PWS Publishers, 1994) 166.

187 *The introduction of chaos theory:* For the story of the development of chaos theory and its mixed initial reception, see generally Gleick, *Chaos.*

187 *To say that either politics:* See Ian Percival, "Chaos: A Science for the Real World," in *Exploring Chaos,* ed. Nina Hall, (New York: Norton, 1991), 16.

187 *One major characteristic of chaotic systems:* Robert L. Devaney, *Introduction to Chaotic Dynamical Systems* (2d ed.; Redwood City, Calif.: Addison-Wesley, 1989), 2.

188 *"sensitive dependence on initial conditions":* Ibid, 50.

188 *"[T]he behavior of systems with different initial conditions":* Peter Coveney, "Chaos, Entropy and the Arrow of Time," in *Exploring Chaos,* Hall, 203, 210.

188 *topologically transitive:* Devaney, *Introduction to Chaotic Dynamical Systems,* 50.

188 *Regardless of the initial density:* If $a$ is less than 1, the population eventu-

ally dies out; that is, it settles at 0. For *a* between 1 and 3, the population eventually settles at some fixed density (Robert May, "The Chaotic Rhythms of Life," in *Exploring Chaos,* Hall, 82, 83–84).

190 *complicated results do not necessarily imply:* See Paul Davies, "Is the Universe a Machine?" in *Exploring Chaos,* Hall, 213, 220.

192 *This is sometimes called the butterfly effect:* See John L. Casti, *Complexification* (New York: HarperCollins, 1994), 89.

192 *If you were 100,000 times more accurate:* Franco Vivaldi, "An Experiment with Mathematics," in *Exploring Chaos,* Hall, 41. It may be useful at this point to realize that sufficient precision to cure the problem is not only unachievable in a practical sense, it is theoretically impossible. If we measure two distances, with superhuman accuracy, and achieve the same length, 2.0000000000 inches, both times, we still do not know if they are the same: the correct distance for the first could really be 2.0000000000*1* inches, for the second 2.0000000000*2* inches. There is no stopping this expansion of decimals—to prove that there is no error requires an infinite quantity of information, which is impossible (Davies, "Is the Universe a Machine?" 213, 219).

192 "*We have arrived at the core of the issue*": Vivaldi, "An Experiment with Mathematics," 41.

193 *yet they may still be bounded:* Carl Murray, "Is the Solar System Stable?" in *Exploring Chaos,* Hall, 96, 106.

193 *While the precise direction:* See Ian Stewart, "Portraits of Chaos," in *Exploring Chaos,* Hall, 44, 50–52.

193 *the bifurcations' occurring at shorter and shorter intervals:* Period-doubling bifurcations have been termed a typical route to chaos (Devaney, *Introduction to Chaotic Dynamical Systems,* 130).

193 *at the edge of chaos:* See, for example, Stuart A. Kauffman "Antichaos and Adaptation," *Scientific American,* August 1991, 78–84.

194 *coherence under change:* See John H. Holland, *Hidden Order* (Reading, Mass.: Addison-Wesley, 1995), 4.

194 *complex adaptive systems:* Ibid., 2–4.

194 *The most successful systems:* See, for example, Kauffman, "Antichaos and Adaptation," 78–84.

194 "*just the right balance*": M. Mitchell Waldrop, *Complexity* (New York: Simon & Schuster, 1992), 308.

194 "*Highly chaotic networks*": Kauffman, "Antichaos and Adaptation," 78–84. This phase transition is an imprecise analogy, as the liquid phase is not a mere transition but its own distinct phase of matter (ibid.).

194 *a discontinuous shift:* See Casti, *Complexification,* 53.

195 *Inordinate hoopla surrounded catastrophe theory:* For an account of the rise and fall of catastrophe theory, see John Horgan, *The End of Science* (Reading, Mass.: Addison-Wesley Longman, 1996), 208; John Casti, *Searching for Certainty* (New York: Morrow, 1990), 63–64.

195 *catastrophe theory provides interesting* illustrations: See Casti, *Complexification,* 47–48.

196 *"Clouds are not spheres":* Benoit Mandelbrot, "Fractals—A Geometry of Nature," in *Exploring Chaos,* Hall, 122, 125. See generally Benoit Mandelbrot, *The Fractal Geometry of Nature* (2d ed; San Francisco: W. H. Freeman, 1982).

196 *"Speaking casually":* Rucker, *Mind Tools,* 176.

197 *the theory of chaos offers:* Some important recent articles in this area include J. B. Ruhl and Harold J. Ruhl Jr., "The Arrow of the Law In Modern Administrative States: Using Complexity Theory to Reveal the Diminishing Returns and Increasing Risks the Burgeoning of Law Poses to Society," *University of California at Davis Law Review* 30 (1997): 405; Glen Harlan Reynolds, "Is Democracy Like Sex?" *Vanderbilt Law Review* 48 (1995): 1635; Thomas Earl Geu, "The Tao of Jurisprudence: Chaos, Brain Science, Synchronicity, and the Law," *Tennessee Law Review* 61 (1994): 933; Andrew W. Hayes, "An Introduction to Chaos and the Law," *University of Missouri–Kansas City Law Review* 60 (1992): 751; Mark J. Roe, "Chaos and Evolution in Law and Economics," *Harvard Law Review* 109 (1996): 641; Robert E. Scott, "Chaos Theory and the Justice Paradox," *William and Mary Law Review* 35 (1993): 329; Michael J. Gerhardt, "The Role of Precedent in Constitutional Decisionmaking and Theory," *George Washington Law Review* 60 (1991): 68; Glenn Harlan Reynolds, "Chaos and the Court," *Columbia Law Review* 91 (1991): 110; William H. Rodgers, "Where Environmental Law and Biology Meet: Of Pandas' Thumbs, Statutory Sleepers, and Effective Law," *Univerity of Colorado Law Review* 65 (1993): 25; Edward S. Adams et al., "At the End of Palsgraf, There is Chaos: An Assessment of Proximate Cause in Light of Chaos Theory," *University of Pittsburg Law Review* 59 (1998): 507; Nancy Levit, "Symposium on the Trends in Legal Citations and Scholarship: Defining Cutting Edge Scholarship: Feminism and Criteria of Rationality," *Chicago-Kent Law Review* 71 (1996): 947; Vincent M. Di Lorenzo, "Equal Economic Opportunity: Corporate Social Responsibility in the New Millennium," *University of Colorado Law Review* 71 (2000): 51; Susan Bandes, "*Terry v. Ohio* in Hindsight: The Perils of Predicting the Past," *Constitutional Commentary* 16 (1999): 491.

197 *a line of cases:* For example, *Robinson v. Neil,* 409 U.S. 505, 508 (1973).

197 *a pattern of decisions:* For example, *U.S. v. Johnson,* 457 U.S. 537, 564 (1982) (Justice White dissenting).

197 *"The drawback in using a fractal curve":* Rucker, *Mind Tools,* 174–75. See also Mandelbrot, "Fractals—A Geometry of Nature," 122, 132–33 (stating that "[T]the fractal description breaks down on the very small scale and on the very large. Trees or arteries do not branch endlessly, and whole trees are not part of supertrees").

198 *The 1922 case granting a Fifth Amendment right:* See *Pennsylvania Coal v. Mahon,* 260 U.S. 393 (1922); *Lucas v. South Carolina Coastal Council,* 505 U.S. 1003 (1992); *Dolan v. City of Tigard,* 512 U.S. 374 (1994).

198 *Similarly, the Court's striking down of a law barring parents:* See *Pierce v. Society of Sisters,* 268 U.S. 510 (1925); *Griswold v. Connecticut,* 381 U.S. 479 (1965); *Roe v. Wade,* 410 U.S. 113 (1975).

198 *"How can simple rules":* Rucker, *Mind Tools,* 314.

199 *"because the alternatives are binary:* Richard Kay, "Adherence to the Original Intentions in Constitutional Adjudication: Three Objections and Responses," *Northwestern University Law Review* 82 (1988): 226, 243–44 (emphasis in original). There are numerous reasons that the determination of original intent is difficult. Not only is it never a trivial matter to discover intent and understanding, especially two centuries after the fact, but the problem is compounded by the fact that there were many different individuals, who had multiple motives and understandings, involved.

199 *"the smallest change leads":* David Tritton, "Chaos in the Swing of a Pendulum," in *Exploring Chaos,* Hall, 22, 28.

199 *"[I]f you consider the evolution of doctrines":* Paul Brest, "The Misconceived Quest for the Original Understanding," *Boston University Law Review* 20 (1980): 204, 234.

200 *"It is a maxim among these lawyers":* Jonathan Swift, *Gulliver's Travels* (1726; New York: Gramercy Books, 1995), 241.

200 *"duty to reconsider constitutional interpretations": Planned Parenthood of Southeastern Pennsylvania v. Casey,* 505 U.S. 833, 954 (1992) (Chief Justice Rehnquist dissenting). See also *Burnet v. Coronado Oil & Gas Co.,* 285 U.S. 393, 406 (Justice Brandeis dissenting) (describing the need to overrule erroneous constitutional cases, because "correction through legislative process is practically impossible").

200 *"serious inequity to those who have relied":* See *Casey,* 505 U.S. 855; *Moragne v. States Marine Lines,* 398 U.S. 375, 403 (1970). See generally Henry Paul Monagham, "Stare Decisis and Constitutional Adjudication," *Columbia Law Review* 88 (1988): 723, 748 (discussing the values of adherence to

precedence, as "consistency, coherence, fairness, equality, predictability, and efficiency").

201 *"in a chaotic system errors grow":* Davies, "Is the Universe a Machine?" 213, 219.

201 *constitutional moment:* See, for example, Ackerman and Golove, "Is NAFTA Unconstitutional?" 801, 873.

201 *"The 1937 watershed marked":* Tribe, "Taking Text and Structure Seriously," 1296–97. For an example of this change, compare *N.L.R.B. v. Jones and Laughlin Steel Corp.,* 301 U.S. 1 (1937) (upholding federal regulation of manufacturing) with *Carter v. Carter Coal Co.,* 298 U.S. 238 (1935) (striking down federal regulation of mining because it involved "production, not . . . trade").

201 *"forecloses us from reverting": United States v. Lopez,* 514 U.S. 549, 574 (Justice Kennedy concurring).

201 *jump discontinuity:* Casti, *Complexification,* 46.

203 *"An interpretation is legitimate":* Frederick Schauer, "Easy Cases," *Southern California Law Review* 58 (1985): 399, 431. See also Kay, "Adherence to the Original Intentions in Constitutional Adjudication," 249 (stating that "By discerning the language's central paradigm, we can define an area of application that was intended by virtually all the relevant individuals . . .").

203 *"The Court is most vulnerable": Bowers v. Hardwick,* 478 U.S. 186, 194 (1986). Of course, there can still be dispute over what are the "roots in the language or design of the Constitution." In *Bowers,* Justice White stated for the Court that "to claim that a right to engage in [consensual sodomy] is 'deeply rooted in this Nation's history and tradition' or 'implicit in the concept of ordered liberty' is, at best, facetious." Justice Blackmun, in dissent, argued that, "depriving individuals of the right to choose for themselves how to conduct their intimate relationships poses a far greater threat to the values most deeply rooted in our Nation's history than tolerance could ever do." Ibid., 214 (Justice Blackmun dissenting).

203 *the system of appointment:* U.S. Const., Art. II, §2.

203 *a helix resembles an endless Slinky:* See, for example, Rucker, *Mind Tools,* 155. The helix that looks like it is wrapped around a cylinder is more precisely known as a circular helix.

204 *Human intellectual development:* Ibid.

205 *coherence under change:* See John H. Holland, *Hidden Order* (1995), 4.

205 *balance of stability and fluidity:* Waldrop, *Complexity,* 308. For interesting legal applications of complexity, see J. B. Ruhl, "The Fitness of

Law: Using Complexity Theory to Describe the Evolution of Law and Society and Its Practical Meaning for Democracy," *Vanderbilt Law Review* 49 (1996): 1407; J. B. Ruhl, "Complexity Theory as a Paradigm for the Dynamical Law-and-Society System: A Wake-up Call for Legal Reductionism and the Modern Administrative State," *Duke Law Journal* 45 (1996): 849; Gerald Andrews Emison, "The Potential for Unconventional Progress: Complex Adaptive Systems and Environmental Quality Policy," *Duke Environmental Law and Policy Forum* 7 (1997): 167.

205 "*whenever it is executed contrary to their Interest*": Quoted in *U.S. Term Limits, Inc. v. Thornton,* 514 U.S. 779, 814 n. 26 (1995).

205 "*[f]requent elections are unquestionably*": Federalist No. 52 (Hamilton or Madison).

205 "*The mutability in the public councils*": Federalist No. 62 (Hamilton or Madison).

205 *The members of the two houses of the legislature:* U.S. Const., Art. I, § 2, cl. 1 and § 3, cls. 1 and 2.

206 Thornton: 514 U.S. 779 (1995).

206 "*entrenched incumbency*": Preamble to Amendment 73 (Term Limitation Amendment), quoted in *Thornton,* 514 U.S. 784.

206 "*as often as the electors shall think fit*": A Citizen of New Haven, "Observations on the New Federal Constitution," *Connecticut Currant,* January 7, 1788; reprinted in Kaminski and Saldino, *Documentary History,* vol. 15, 280.

207 "*in no way freezes the status quo*": *Jenness v. Fortson,* 403 U.S. 431, 439 (1971).

207 Williams: 393 U.S. 23, 24–25 (1968).

207 "*Competition in ideas and governmental policies*": *Williams v. Rhodes,* 393 U.S. 31–32.

207 Timmons: 520 U.S. 351 (1997).

208 *bootstrap*": Ibid., 366.

208 "*a strong interest in the stability*": Ibid., 367.

208 "*the two old, established parties*": *Williams v. Rhodes,* 393 U.S. 31. Similarly, in *Anderson v. Celebrezze,* 460 U.S. 780 (1983), the Court struck down Ohio's law requiring independent presidential candidates to file nominating petitions in March in order to appear on the election ballot in November. The Court rejected Ohio's assertion that such early filing furthered its interest in "political stability," because "the asserted interest in political stability amounts to a desire to protect existing political parties from competition . . ." (ibid., 801–2).

208 *"it is plain . . . that realistic accounts":* David Berlinski, *On Systems Analysis* (1976), 137.

208 *"Not only in research, but also in the everyday world":* May, "Simple Mathematical Models with Very Complicated Dynamics," 467.

## 11. The Mathematics of Limits

209 *"But a constitution is framed for ages": Cohens v. Virginia,* 19 U.S. 264, 387 (1821).

209 *tells the story:* This story is adapted from David Berlinski, *A Tour of the Calculus* (New York: Pantheon Books, 1995), 135.

210 *you can't divide by 0:* One reason you can't divide by 0 is because it would destroy the symmetry with multiplication. For example, $6/3 = 2$ is the same as $2 \times 3 = 6$. Put algebraically, $A/B = C$, means $C \times B = A$. But if you take a fraction like $6/0$, you're in trouble, since anything multiplied by 0 equals 0. So if you said that $6/0 = \infty$, that would have to mean that $\infty \times 0 = 6$. Or if $6/0 = 0$, then $0 \times 0 = 6$. Since none of these make sense, mathematicians just prohibit division by 0.

211 *The trick is to take smaller and smaller time intervals:* This example is derived from Kline, *Mathematics in Western Culture,* 217–18.

211 *"is not defined as the quotient":* Ibid., 218.

211 *Limits also are critically important:* Limits are also essential for understanding calculus. See, for example, Berlinski, *A Tour of the Calculus,* 164–69.

211 *A limit in mathematics has a very rigorous, very formidable definition:* If $L$ is the limit of $f(x)$, we say that
$\lim_{x \to a} f(x) = L$ if given any $\varepsilon > 0$, there is a $\delta > 0$ such that $|f(x) - L| < \varepsilon$, whenever $0 < |x - a| < \delta$.

211 *"our intuition of something tending":* Paulos, *Beyond Numeracy,* 130.

211 "*The* concept *of a limit is simple":* Berlinski, *A Tour of the Calculus,* 120.

212 *focusing on the weaknesses faced by the country:* Story, *Commentaries on the Constitution,* secs. 225–35.

212 *"All acknowledge that they were convened": Cohens,* 19 U.S. 416–17. See also *Chisholm v. Georgia,* 2 U.S. 419, 465 (1793):

> *One of [the Constitution's] declared objects is, to form an union more perfect, than, before that time, had been formed. Before that time, the Union possessed Legislative, but uninforced Legislative power over the States.*

*Nothing could be more natural than to intend that this Legislative power should be enforced by powers Executive and Judicial.*

The holding of *Chisholm,* permitting lawsuits against individual states without their consent, was overruled by the Eleventh Amendment.

213 *"would be unable to govern a nation":* Robert Shoemaker, "Democracy and Republic as Understood in Late Eighteenth Century America," *American Speech* 41 (1966): 83, 95 (emphasis added).

213 *"[I]t is a sufficient recommendation of the Federal Constitution":* Federalist No. 43 (Madison).

213 *"basic human rights of members of groups": R.A.V. v. City of St. Paul,* 505 U.S. 377, 395 (1992).

214 *"the remedy to be applied": Whitney v. California,* 274 U.S. 357, 377 (1927) (Justice Brandeis concurring).

214 *"And though all the winds of doctrine were let loose":* John Milton, *Areopagitica* (1644).

214 *This series is actually a very slow way:* Maor, *To Infinity and Beyond,* 34-36.

214 *"In the long run, true ideas do tend to drive out":* Wellington, "On Freedom of Expression," *Yale Law Journal* 88 (1979): 1105, 1130–32.

215 *"realistic possibility that official suppression": R.A.V.,* 505 U.S. 390.

215 *"[t]he least prejudicial sellers":* Posner, *Economic Analysis of Law,* 625.

215 *There is some debate:* Compare Richard A. Epstein, "The Status-Production Sideshow: Why the Anti-Discrimination Laws are Still a Mistake," *Harvard Law Review* 108 (1995): 1085, with Richard H. McAdams, "Cooperation and Conflict: The Economics of Group Status Production and Race Discrimination," ibid., 1003.

216 *"But when you have seen":* Martin Luther King Jr., "Letter from Birmingham Jail," recently reprinted in *University of California at Davis Law Review* 26 (1993): 835, 839–40. It can also be found in *Why We Can't Wait,* Martin Luther King Jr. (New York: Harper & Row, 1964), 77, 83–84.

## 12. The Limits of Mathematics

217 *"As far as the laws of mathematics refer":* Albert Einstein, "Geometry and Experience," in *Sidelights on Relativity,* trans. G. B. Jeffrey and W. Perrett (1923; reprint, New York: Dover, 1983), 25, 28.

217 *"Sam, if a man can walk":* E. B. White, *The Trumpet of the Swan* (New York: HarperCollins, 1970), 63–64.

218 *"wishing to find laws for cloud motion":* Hofstadter, *Fluid Concepts and Creative Analogies,* 125.

219 *"infinitely complex and impossible to capture completely":* Paulos, *Beyond Numeracy,* 149. The quoted passage describes reality in general but certainly applies to the subset of constitutional reality as well.

219 *we should not look to mathematics:* Compare Laurence Tribe, "The Curvature of Constitutional Space: What Lawyers Can Learn from Modern Physics," *Harvard Law Review* 103 (1989): 1, 2 (warning of the futility of "searching the sciences looking for authoritative answers to legal questions").

219 *"Quantitative work has its special dangers":* Berlinski, *On Systems Analysis,* 48.

219 *It is simply inappropriate to present DNA evidence:* See, for example, Tribe, "Trial by Mathematics," 1329, 1365, asking how a jury could be expected to balance impressive numbers "against such fuzzy imponderables as the risk of frame-up or of misobservation, if indeed it is not induced to ignore those imponderables altogether."

219 *"Mathematics, a veritable sorcerer": People v. Collins,* 438 P. 2d 33, 33 (Cal. 1968).

220 *"All is number":* See, for example, Paulos, *Beyond Numeracy,* 193; Kline, *Mathematics in Western Culture,* 40.

220 *"Let us calculate":* Rucker, *Mind Tools,* 197. See also Kline, *Mathematics in Western Culture,* 239–40. Similarly, the eighteenth-century French mathematician Marie Jean Antoine Nicolas de Caritat, Marquis de Condorcet, strove to create a system of "social mathematics." See generally Baker, *Condorcet.* Robespierre summed up the contemporaneous view of Condorcet and his philosophy in the damning phrase: "[H]e was a great mathematician in the eyes of men of letters, and a distinguished man of letters in the eyes of the mathematicians" (ibid., 383).

220 *For students of the Constitution:* See, for example, Dow, "Gödel and Langdell," 726: "Law is norms, and normative advances depend largely on nonmathematical intelligence."

220 *But the choice between competing values:* See, for example, Paulos, *Beyond Numeracy,* 261: "In matters of social policy or personal decision making, mathematics can help determine the consequences of our assumptions and values, but it is we (we X's), not some mathematical divinities, who are the origin of these assumptions and values." See also Dow, "Gödel and Langdell," 710–11 (stating that "analogies between law and mathematics . . . tend to perpetuate the canard that law is like

science or mathematics, when in fact it is not. What law is really like is religion . . .").

220 *No computer can be programmed:* See, for example, Tribe and Dorf, *On Reading the Constitution,* 96: "[L]ogical consistency is too weak a condition to discriminate between competing abstractions."

221 *"Mathematics never captures all":* Devlin, *Mathematics,* 71.

# Permissions

Grateful acknowledgment is made to the following for permission to reprint copyrighted materials:

Page 41 Graphs from *Clinical Epidemiology and Biostatistics* by Rebecca Knapp and M. Clinton Miller. Copyright © 1992 Lippincott Williams & Wilkins. Used by permission of Lippincott Williams & Wilkins.

Page 135 Figure from *Concepts of Modern Mathematics* by I. Stewart. Penguin Books, Ltd. 1975. Reprinted by permission of the author.

Page 136 Figure from *Mathematics: The Science of Patterns* by Keith Devlin. Copyright © 1997 by Keith Devlin. Reprinted by permission of Henry Holt and Company, LLC.

Page 139 Figure from *Archimedes' Revenge* by Paul Hoffman. Copyright © 1988 by Paul Hoffman. Used by permission of W. W. Norton & Company, Inc.

Page 147 "Outwitted" by Edwin Markam from *Best Loved Poems of the American People* (1957) c. 1936, originally published by Doubleday, now owned by Random House. Public Domain.

Page 189 Graphs from "When two and two do not make four: non-linear phenomena in ecology" by Robert May, *Proceedings of the Royal Society of London* (1986). Copyright © 1996 by Robert May. Reprinted by permission of Lord May of Oxford and the Royal Society.

Page 190 Charts from *Exploring Chaos: A Guide to the New Science of Disorder,* edited by Nina Hall. Copyright © 1991 by IPC Magazines New Scientist. Used by permission of W. W. Norton & Company, Inc.

Page 195 Figure from *The Fractal Geometry of Nature* by Benoit B. Mandelbrot, published by W. H. Freeman. Copyright © 1982 by Benoit B. Mandelbrot. Reprinted with permission.

Page 202 Figure adapted from *Complexification* by John L. Casti. Copyright © 1994 by John L. Casti. Reprinted by permission of HarperCollins Publishers, Inc.

Page 209 Excerpt from *A Tour of Calculus* by David Berlinski. Copyright © 1995 by David Berlinski. Used by permission of Pantheon Books, a division of Random House, Inc.

Page 216 Excerpt from letter written by Martin Luther King Jr. reprinted by arrangement with the Estate of Martin Luther King Jr., c/o Writers House as agents for the author.

Page 217 Excerpt from *The Trumpet of the Swan* by E. B. White. Copyright © 1970 by E. B. White. Reprinted by permission of HarperCollins Publisher Inc.

# Index